Praise for the book

Aadhaar is a sapling that I once nurtured. This book refreshes the memory of its growth in an entertaining manner. It is an authentic, gripping and meticulous account of how Aadhaar came to be. The making of what is now the world's largest identification platform holds lessons for leaders across domains and geographies. From its audacity to its methodical implementation to its manifold applications, Aadhaar will be remembered as one of the most important interventions of our times. RS Sharma, the man steering the mission, has done a fabulous job of compiling vital lessons which will serve as a manifesto for everyone who hopes to harness technology for the betterment of humanity.

—**Pranab Mukherjee**, former President of India

Many times, when I was despondent about whether we would succeed, I would be so thankful for my luck and good fortune that Ram Sewak was our CEO. I have no doubt that the project would have failed miserably but for him. He was truly the Aadhaar of Aadhaar!

The book is a very articulate, gripping and candid account of the story of Aadhaar.

—**Nandan Nilekani**, founding chairman, UIDAI

Cynics, and there were many, said it just can't be done. The Aadhaar team proved them wrong, and wow how awfully wrong! *The Making of Aadhaar* is a remarkable success story told by a remarkable storyteller.

—**D. Subbarao**, former governor, Reserve Bank of India

This book is a saga of the buccaneering voyage of the Aadhaar ship braving uncharted seas, Eddystone Rocks of administrative impedance and rights activists, and sailing into safe harbour.

—**Justice B.N. Srikrishna**, Former Judge of the Supreme Court of India, Chairman of FSLRC and Chairman of the Committee on Data Protection Law

My distinguished colleague Dr Ram Sewak Sharma has laid out several key ideas with the precision of the mathematician that he is. And like in all mathematics, ideas in this book presage the world of tomorrow. His wonderful contribution to bring India's information economy-related ecosystem to the cutting edge provides a platform to our young generation to launch the next wave of productivity-led growth. I warmly recommend this book to all our aspirational fellow citizens.

—**Vijay Laxman Kelkar**, chairman, India Development Foundation

THE MAKING OF
AADHAAR

THE MAKING OF
AADHAAR
World's Largest Identity Platform

Ram Sewak Sharma

RUPA

Published by
Rupa Publications India Pvt. Ltd 2020
7/16, Ansari Road, Daryaganj
New Delhi 110002

Sales centres:
Allahabad Bengaluru Chennai
Hyderabad Jaipur Kathmandu
Kolkata Mumbai

The views and opinions expressed in this book are
the author's own and the facts are as reported by him
which have been verified to the extent possible,
and the publishers are not in any way liable for the same.

ISBN: 978-93-90356-12-6

First impression 2020

10 9 8 7 6 5 4 3 2 1

The moral right of the author has been asserted.

CONTENTS

CONTENTS

FOREWORD

In my first book, *Imagining India: Ideas for the New Century*, I had written about the need for a unique identification platform to fix the country's leaky welfare distribution infrastructure. The government had, in parallel, thought of a unique ID system which was approved by the Cabinet in early 2009. My book gave it the confidence that I could do the job. Dr Manmohan Singh (then prime minister) gave me a chance to pursue this mission.

In 2009, I was invited to join the government to plan and implement the project. I joined the Unique Identification Authority of India (UIDAI) as its first chairman. Even with a limited experience of the internal workings of the government—thanks to my involvement with the Bangalore Agenda Task Force (BATF)—under the leadership of Mr S.M. Krishna, the former chief minister (CM) of Karnataka—I knew that a task as ambitious as giving a unique ID to all Indian residents would need an amazing CEO, ideally a maverick bureaucrat familiar with administrative processes and open to the transformative power of technology. I shared this challenging job description with a friend in the Indian Administrative Service (IAS), K.P. Krishnan. KP, prompt as ever, made an interesting introduction.

Enter Ram Sewak Sharma, the personification of all the attributes that were essential for our mission to succeed—a career mandarin with healthy disregard for convention, a versatile administrator in an obsessive relationship with code, a tough taskmaster with a profound empathy for the human condition. When I shared the genesis of Aadhaar with Ram Sewak, our impending partnership seemed divinely preordained rather than a fortuitous coincidence. Irrespective of the nature of his postings, Ram Sewak had spent his entire career in introducing efficiencies into governance using software. In each of his assignments—as Joint Secretary in Bihar's Irrigation department; as District Magistrate, Begusarai, Purnea and Dhanbad; as Bihar's State Transport Commissioner, in the state's treasury and at the Central Department of Economic Affairs, Ram Sewak had put his technical skills to good use. He wrote programmes to reduce various kinds of friction. With an earnest belief that technology could rid India of several ailments, Ram Sewak supplemented his interest with a master's degree in computer science from the University of California, Riverside (UCR) before returning to fulfil his mandates with greater gusto. When I offered him the role of Mission Director and Director General (DG) of UIDAI, Ram Sewak's acceptance was visceral.

My earliest memory of our working relationship is from a shared, makeshift workspace in the erstwhile Yojana Bhawan (now NITI Aayog). Coming from the informal working culture of Bangalore (now Bengaluru), the setting was fine by me, but a downgrade for Ram Sewak, who must have certainly been used to more elaborate arrangements. However, his easy adaptability gave me the early confidence in his ability to improvise—a skill that would come in handy several times over the next four years of our working relationship. The scrappiness was also an omen of things to come. Given the competing demands of public administration in India, we would often have to do justice to a gargantuan goal with limited resources. Since the organization was thinly staffed at that point, we also got the opportunity to develop mutual trust that enabled frequent tag teaming. I often look back at the early guidance that Ram Sewak and Srikar M.S. (my private secretary from the IAS) proffered as the enabler of Aadhaar's successes.

Ram Sewak was a patient guide who eased me into the ways of the government. When I argued for a reduction in the number of sanctioned posts (molded as I was in corporate principles such as leanness), Ram Sewak advised me to avoid doing so explicitly. He reasoned that we could always leave certain posts unfilled. He did this with quintessential candor when he said, 'You could go for the reduction, but you are not going to get a Param Vir Chakra for this act!' Such pithiness will be familiar to those who know Ram Sewak. Many times, when I was despondent about whether we would succeed, I would be so thankful for my luck and good fortune that Ram Sewak was our CEO. I have no doubt that the project would have failed miserably but for him. He was truly the Aadhaar of Aadhaar!

Ram Sewak was also a great listener. When I suggested that we recruit officers from across different administrative services—an idea that could have easily been misinterpreted as an affront to the supremacy of the IAS—he immediately agreed and we went on to assemble some of the finest minds from diverse arms of the government. My subtle hints on valuing ability over experience— the inverse of governmental convention—were also well taken.

It is largely due to Ram Sewak that we could inject lateral thinking into our assignment. He was magnanimous when it came to receiving and accommodating private-sector professionals. That said, his generosity only extended up to a point, and with reason. When some of our colleagues tried to project that the private sector had played a disproportionate role in making Aadhaar happen, Ram Sewak objected. The bureaucracy was particularly hurt by accounts that made it seem redundant. This view that the heavy-lifting was done by lateral entrants was also patently false. While several team members contributed to the mission, its seamless implementation was due to the public sector's intrinsic ability to pull things off at scale. I had seen evidence of this in the work of the Election Commission and in other mission mode endeavours of the government before I experienced it first hand at UIDAI.

The only real disagreements that Ram Sewak and I had were on the Aadhaar artefact itself and on our approach towards non-governmental outreach. Ram Sewak handled these gracefully. I had always argued for Aadhaar to be a number, not a card. Ram Sewak, more in touch with residents' aspirations, passionately advocated for a card. This division at the top had led to groupism within the organization. It took Ram Sewak one delicate metaphor to settle the matter. He argued that the card was the body, while the number was the soul. Without the number, there would be no card! And with that, we brokered peace.

Given the complexity and scale of our mission, we had constituted a Civil Society Outreach (CSO) arm within the organization. The mandate of this group was to consult with non-governmental organizations (NGOs) across geographies and sectors to understand the challenges that marginalized sections of our population face. Their enrolment was a priority for us. We gained invaluable insights from these consultations that informed and strengthened our design. However, select actors started doubting our intent while entirely disregarding the stated objective. They ran a smear campaign, which hurt Ram Sewak, who had always been apprehensive about engaging with NGOs. I had tried to allay his fears, but his hunch that they will do more harm than good in our limited context proved to be prescient. As Ram Sewak surmises in this book, we became perennial punching bags for them.

Ram Sewak's ability to roll his sleeves up for our common cause knew no bounds. Having suffered the byzantine procurement policies of the government, Ram Sewak knew that relying on the selection and onboarding of our technology partners would invariably delay the kick-off. This was unacceptable since we had made a public commitment to enroll half of the nation by 2014. This tension brought out the irrepressible coder in Ram Sewak. He would wake up early to write code and put together the first enrolment client we used for our proof of concept. He did not stop at the enrolment software. Throughout the mission, he would find time to both review and develop prototypes to ensure that the lag between intent and outcome was minimal. This dedication was both atypical and inspiring for the rank and file of the organization.

While I did capture broad observations and key lessons from my experiences in a book (*Rebooting India: Realizing a Billion Aspirations*) I co-authored, the story of Aadhaar was waiting to be written. It is hard to think of a more authoritative account than this one. By relying on his elephantine memory and elaborate archives, Ram Sewak has painstakingly documented an authentic story of the debates that led to Aadhaar's design, the conversations that improved its implementation and the decisions that made it the world's largest developmental platform and the quickest to onboard a billion 'users'.

Aadhaar's evolution spans several realms. While its existence is a testament to the power of ideas, it is the process of converting a thought into an article of faith that provoked deep technical research, involved psychography, unconventional team building, varied legislation, complex experimentation, a collective imagination and even turf wars (in a place where turf wars are a contact sport)! Drawing this arc chronologically, accurately and humorously is this book's astounding achievement. It is a very articulate, gripping and candid account of the story of Aadhaar.

After moving on from UIDAI, Ram Sewak has been an indefatigable torchbearer for Aadhaar. In various capacities in his cadre (Jharkhand) and in the capital, he has introduced Aadhaar-enabled innovations such as attendance systems, pension transfers, registration platforms, eSign and DigiLocker. The last two are components of the IndiaStack—a set of products that allow governments, businesses, start-ups and developers to solve India's hard problems towards presence-less, paperless and cashless service delivery. At least two other products in the Stack already have Ram Sewak's imprints—e-Aadhaar and electronic Know Your Customer (eKYC). Most recently, Ram Sewak was also part of a high-powered committee comprising government officials and tech leaders, set up by the Prime Minister's Office (PMO), to work on solutions to deal with the Coronavirus crisis. It is due to Aadhaar that the government has been able to initiate direct cash transfers to hundreds of millions of Indians who needed financial support to deal with the consequences of the lockdown.

Aadhaar was preloaded with a few virtues including but not limited to minimalism, resident-centric design, platform thinking and scalable microservices. Over the past few years, these have emerged as cardinal assumptions for several business and governmental innovations.

How and why were we able to build country-scale infrastructure? What was the secret sauce behind making diverse actors work together? How did we harness all criticism to strengthen our resolve? What are the possibilities we sought to unleash? These are just some of the questions that Ram Sewak has succinctly answered via this book. It is for this reason that it is essential reading for anybody looking for a guide on how to reimagine the future. I hope you enjoy reading it as much as I did!

11 May 2020   Nandan Nilekani
                                         Founding Chairman, UIDAI

PREFACE

This book isn't written to defend Aadhaar. It isn't an autobiographical account of my days at the Unique Identification Authority of India (UIDAI) either. Nor is it a declassification of materials previously withheld from the public or a compendium of anecdotes from those heady days! One learns from experience. In this book, I share an experience; not a small one by any measure. I also share my evaluation of what that experience could tell us.

The story is factual, but its evaluation would depend upon the person doing the interpretation. Your thoughts may differ from my own and will likely illuminate aspects that I missed. I hope you share them: in a tweet, an email, a blog post, a newspaper op-ed, or maybe even in a book. I hope to learn from all of that.

I've also learnt through experience that social media isn't conducive to a thoughtful exploration of ideas—especially when the topic has already polarized the views of people. We tend to react brutally to a part narrative or argument without taking the opportunity to first understand the context in which it is framed.

To an extent, the nature of the new medium itself is to blame for this fragmented expression and understanding. We know that human beings tend to persist with their viewpoint even in the face of evidence to the contrary. When we don't have the opportunity to reflect in silence, there is even less hope of escaping the 'echo chambers' of our minds.

In a sense, this book is a product of the need for a conversation that's not suited for social media, government reports, newspaper columns or television debates. We need this conversation to capture what made the unique ID project successful, where similar attempts have failed in the past, in India as well as elsewhere.

What made it happen? Was it the leadership of Nandan Nilekani? Was it the collective excitement of doing something new? Was it a unique organizational structure? Did the criticism from all sides serve to strengthen the resolve of the team? And finally, was the lack of precedent a challenge, an opportunity or a curse?

In my entire career with the government spanning more than four decades, I have never worked on a project where each member of the team was so full of passion and commitment. The excitement and involvement of the people was truly remarkable.

The received wisdom is that projects suffer from time and cost overruns, especially those owned by governments. But here was a project that was run in a hybrid mode, with government and private-sector participation, which decided in favour of a technology that hadn't been proven at the required scale, thus overcoming opposition from several quarters and delivering within aggressive cost and time targets. It is probably the only project where cost estimates for some activities were revised downwards.

The project would be defined a success because more than 1.25 billion Indian residents have been enrolled into Aadhaar, 42 billion authentications have been carried out and 8.2 billion eKYCs have been made possible till date.[1] A comparable success story in numbers would be difficult to find.

If you don't agree with this definition of success, you can measure Aadhaar with your own definition. What would it be? Speed of execution of the project? Savings to the government? Accuracy in deduplication or authentication in an ID project? Convenience to the public? Or the fact that Aadhaar came through in a challenge before the Supreme Court that led to a recognition of constitutional protection to the right to privacy?

You may agree or disagree with the premise that a unique identity for each resident is required. You may conceive of solutions that work differently from the way Aadhaar does. But I expect that you are not entirely dismissive of what has been accomplished in this project and would like success similarly replicated in other initiatives of the government or the private sector as well. In that case, this book is for you.

---

[1] Updated data is available on the Aadhaar dashboard of Unique Identification Authority of India (UIDAI) at: https://uidai.gov.in/aadhaar_dashboard/. Last accessed on 20 July 2020.

Stories are the most persuasive tools known to man. And therein lies the trap called the narrative fallacy. If you think that Aadhaar happened because of a few decisions, good or bad, you have already shut out of your mind the myriad ways in which it could have failed.

So, to what do we owe its success? As with most innovative achievements, the answer is layered, but compelling. A constellation of forces aligned to create this momentous story. For good luck needed in future projects, we can only pray to the gods and break the ceremonial coconut. But success that is attributable to foresight, skill or serendipity can surely be captured as useful learning.

I hope that as you read the book, you don't come away thinking it as a self-congratulatory tome. It is offered as a case study, even if my own closeness to the project prevents me from being objective. If this story could even marginally increase the probability of success of other projects, from the smallest personal efforts to large government undertakings, I should consider the effort compensated in full measure.

INTRODUCTION
RITE OF PASSAGE

Before joining the Indian Administrative Service in 1978, I was a student of mathematics at the Indian Institute of Technology (IIT), Kanpur. We had a mainframe at the institute, housed in its own building, which beckoned the child in me to step within and explore. The ever-blinking lights and automatic readers that swallowed cards seemed straight out of a science-fiction movie. It was a state-of-the-art mainframe that only the privileged could see at one of India's premier technology institutes. Little did I know then that fate would keep me close to computers and information systems.

On joining the Service, I was allotted the Bihar cadre and initially posted at various small places such as Chaibasa, Godda and Saharsa.

In May 1983, I was posted to the Secretariat in Patna as Joint Secretary in the Department of Irrigation. One of my responsibilities was looking after the transfers, postings, promotions, etc., of about 4,000 Assistant Engineers (AEs) in the department, who worked in the field and in offices spread across the state. There were two broad categories of work that AEs performed. The first related to preparing designs, estimates and project reports. For some reason, it was called 'Non-works'. The other related to construction and implementation of projects, which was, unsurprisingly, called 'Works'.

The AEs generally preferred the Works. There was a constant tussle for migration from Non-works to Works and for continuing in Works for as long as one could. There was a rule to alternate every three years from Works to Non-works and vice versa, but political and other influence would be brought in to intercede on behalf of some candidates for being posted to or being continued in the Works category.

Organizing the Chaos

The records of the AEs' service history were an absolute mess. Manually managing a cadre of 4,000 personnel without any credible data relating to their employment history was a difficult task. Some continued to stay in their preferred position long beyond the stipulated period of three years, while others languished in less-favoured ones. Rogues could even get the records destroyed! As each correspondence started on a new file, influence peddlers had a field day manoeuvring it to any destination of their choice.

The only way to introduce some system in this chaos was to organize the posting records of the AEs. 'Database' wasn't a term in vogue back then, so we prepared a list of the AEs in big registers, one for each jurisdiction of a Chief Engineer (CE). We assigned a five-digit code for each AE's post. In this code, the first two digits stood for the CE, the next one for the superintending engineer (SE), the fourth one for the executive engineer (EE) and the last one for the AE. Once the list was current, we introduced a system of recording transfer notifications in these registers. While the *pairavis* (recommendations) could not be entirely stopped, these registers ensured that we had a strong argument for rejecting an unjust request based on previous posting records of AEs being in violation of the general rotational rule of three years each in Works and Non-works.

While a simple step, it proved useful in keeping information organized in a systematic manner. Senior officers too were saved from the onslaught of *pairavis*. This system, I am told, continued for quite some time after I left the Irrigation department and went on to join as District Magistrate (DM) of Begusarai in Bihar in January 1985.

To the east of Patna lies the district and the eponymous town of Begusarai, earlier a part of Monghyr district. The region, endowed with several rivers, has an agrarian economy. However, it was more famous for its refinery, fertilizer factory and Barauni, a place in Begusarai district known as the 'Moscow of Bihar' for being a stronghold of the Communist Party of India (CPI). Begusarai was also known as a difficult police district due to its high crime rate.

Here, I found myself in the company of my friend and classmate from IIT Kanpur, Arvind Verma, who was posted as Superintendent of Police (SP). As I had a fascination for electronic gadgets, I purchased an electronic typewriter and a computer when I reached there. The electronic typewriter would produce neat and evenly aligned letters and documents. There was no need for the whitener anymore, as the typewriter had a memory store, where one could correct any mistakes and reprint the letter. I was also the first officer in the government to purchase a computer. It was a DCM-Tandy machine that had no hard drive, but used big floppy disks for storage. It had a royal 64KB of memory and used an operating system called CP/M.

Arvind and I discussed the potential use of this gadget. I studied the computer manual and learnt the programming language. It was called dBase, which was an interpreted programming language and database rolled into one. I started to write code and the first application was a listing of lost-and-found firearms with which we hoped to match firearms and connect them with criminal incidents to solve cases.

Matching Lost-and-Found Firearms: Gun ownership entails the risk of firearms being looted or stolen and then used for settling a dispute. Looting of firearms during a dacoity and their subsequent recovery were routine in the crime statistics of our state. There was, and continues till today, a system of record-keeping of these events of loss and recovery of firearms.

The standard operating procedure (SOP) was that when a firearm was lost or looted in any incident, its type, make and serial number would be flashed by the police through the wireless system to all the districts of the state. When any firearms were recovered, whether involved in an incident or not, their details were also shared with all the districts. Such exchange of information was, of course, in the hope that the districts would check if any of these arms belonged to their jurisdiction and whether it could be restored to their rightful owners or also help in solving crimes that involved the use of a gun. Details of these lost-and-found firearms, when flashed, were duly recorded into a register. That is where the information rested because firearms have long alphanumeric serial numbers that are difficult to match manually. Arvind and I thought if we entered the contents of this register in a computer database and indexed it on firearm type with the serial number, the matching ones would be listed together.

When we did just this, lo and behold, there were as many as 22 matching pairs found! This meant that we had solved 22 cases where the firearms could be returned to their rightful owners and a link between two crimes that featured the same firearm could be established, thus providing vital clues to investigators. The police headquarters at Patna could scarcely believe the story until we presented the details to them. It became quite a news at that time in police circles.

Summons for Trial: The trial for a crime must be conducted in the presence of the accused. That's the law. Also, any witnesses to the crime must be brought to the court for recording evidence. There was a system in place to maintain a General Register, or GR in short, in which the next date of hearing of each trial was listed, with the names of the accused and the witnesses to be produced during that trial. This list was maintained for each court.

However, to be effective, what was needed in addition to the trial date was the list of persons for each thana (police station), whose presence the officer-in-charge had to ensure at the courts. By computerizing the records, it was easy to produce the second list from the first.

The officer-in-charge now simply had to submit a report to the SP, making him accountable for compliance with the list given to him. The summons could no longer be scuttled that easily!

'Dead' Teachers Transferred: There were thousands of primary and middle-school teachers employed in the district, about whom there were no systematic records. There were teachers who were staying at the same place for a decade or even more and dabbling in politics rather than performing their assigned work. Although posted close to their homes, they would often skip school entirely.

With some effort, it was possible to put all the available information into a database and code the rules of transfer into a computer programme. The orders that the computer programme generated created a flutter in the community. Many were happy, but those with vested interests were critical too, which we fully expected.

However, our records were not clean enough. Therefore, the computerized orders that were issued included the names of some dead teachers too. While commenting on the story, newspaper headlines noted: 'Dead teachers transferred by the computer!'

Logistics for Assembly Elections, 1985

Elections are a massive exercise in logistics in the world's largest democracy. One of the requirements of holding elections is the large number of vehicles (trucks, buses and jeeps) for transporting men and materials to the polling booths, as also for patrolling to maintain law and order. As state governments do not have so many vehicles at their disposal, the law provides for the district administration to requisition private vehicles for this work. The requisitioned vehicles are used during the election period and later paid for usage on fair terms.

However, owners were reluctant to provide their vehicles for election duty, as it disrupted their normal business. Further, they did not receive the compensation in time, which led to complaints of bribery and corruption. Payments were delayed because the vehicles went to different destinations and clerks took time to compute the compensation due to the owners. It would take months, sometimes years even to clear the bills, often selectively and in instalments. To avoid this situation, owners would often keep their vehicles out of service at a loss to themselves.

Begusarai district too requisitioned private vehicles and often seized them on the road. As National Highway 31 (NH 31) passed through the district, we were also expected to help the less fortunately located districts with surplus catch!

We had an Assembly election in 1985. As we could not do without seizing the vehicles, we decided to solve the payment problem. For this, we sought support from the Barauni refinery of the Indian Oil Corporation (IOC) that helped to process payments to the vehicle owners using their own computer resources. We provided them with details about the requisition vehicles such as the registration number, the vehicle type, the duration of detention and the applicable rates. Their computers printed out slips with amounts to be paid to each owner, thus cutting out months of manual processing that had frustrated everyone on previous occasions.

That year, we were able to pay the dues almost immediately, often along with the release of the vehicles. The vehicle owners were delighted and promised to voluntarily bring out their vehicles for the next elections in Begusarai!

Documenting these initiatives, Arvind and I published a booklet titled, *Computers in District Administration*, where we described applications that could be developed to improve the systems at the district level. When I recently stumbled upon a copy, it was interesting to note that the booklet contained ideas that went on to become the central themes and subjects of e-governance later.

When we shared some of this work at the Collectors' Conference or at other meetings with colleagues, the usual question however, was not about the work that was done, but the Rules under which we had purchased the computer!

In May 1986, I was shifted as DM of Purnea in Bihar.

A Breach Is Contained

Purnea had a problem that humanity has had to deal with since ancient times: the control and ownership of land that divides humans into aristocrats and commoners. Bihar was the first state in the country to abolish Zamindari and enact a land-reform law, way back in 1950. The legislation stipulated that each person was entitled to only a few acres of land and the 'surplus' land was to vest in the state. Purnea was infamous for its inequitable land ownership. It had a large number of land-ceiling cases in revenue courts at all levels: subdivisional, district, division and Board of Revenue. Surplus land, after disposal of these cases, was to be distributed among the eligible categories: the poor, the landless and the weaker sections of society.

There were, however, methods fair and foul, to prolong court cases for inordinately long periods. Even when decisions were given, landowners appealed against these verdicts and any distribution of land that was to be done would be cancelled or stayed. Further, due to sloppy maintenance of records coupled with corruption, the same land was often distributed to multiple parties at different points in time. This caused much litigation and sometimes violence, and the poor, landless people never seemed to come in possession of the allotted land.

Unfortunately, there was no system of monitoring these cases, which numbered in thousands, pending in various revenue courts. Creating a database of dispute cases and land distribution seemed to be a fair bet and that's what we sat down to do—collecting information from case records in various courts.

The IBM PC clone in the DM's office was the data centre. It had the most valuable information: land-ownership records and court cases pertaining to land disputes. There was no threat of a data breach because there was no internet and the physical security of a DM's office was no less than that of a data centre. Even if someone had access, few had the skill, elementary as it was, to get the information out. And thereby hangs a tale.

Late Dr Vinayan, founder of Mazdoor Kisan Sangram Samiti (MKSS), an organization of landless labourers, was my friend. Although a non-violent person, he was wanted by the authorities because MKSS was alleged to be the front organization for the Communist Party of India (Marxist–Leninist) [CPI (ML)]—Party Unity, which was categorized as violent.

Dr Vinayan once visited us at Purnea and I told him about the computerization of land records. He took a sample printout, before he left for Patna. From there, he boarded a train to Jehanabad, one of the district towns with agrarian tensions and Left-wing extremism. On the way, the police arrested him at the Gaya railway station.

I was mortified. If the police found that document on his person, it would be simple to deduce where he had got it from. The trail of the data breach would lead straight to the data centre in the DM's room! Fortunately, Dr Vinayan had left the papers at Patna. The breach had been contained.

The computerization of the land-ownership records and the pending court cases, however, paid immediate dividends. It unearthed the long pending cases, in some of which the courts had already given orders. We prioritized those that would yield the largest tracts of surplus land, and allotted portions to eligible categories of people.

Digitizing Public Grievances

Grievances expressed by the public are a fact of daily life in any district. We had more than our fair share in Purnea, especially those related to land disputes. Usually, the complainant would address the grievance to the DM, but mark a copy to the President of India, the Prime Minister (PM), the CM and all the way down the hierarchy to the lowest level of administration, say, the Block Development Officer (BDO).

These applications would reach the district at different points of time. The DM's office would forward the first to the person concerned, say, the executive engineer of the Public Works Department (PWD), and a second copy of the same to another authority for necessary action. When nothing materialized, the aggrieved person would write another application stating that he had already represented in the past.

There was no way of knowing whether action had been taken on the earlier complaint, or if action was taken, was it appropriate and to the satisfaction of the applicant. No one knew how to sort through this mess.

To track the handling of grievances, I decided to write the Public Grievances (PG) software and open a Jan Shikayat Koshang (Public Grievances Cell) in a building of its own. Public grievances were accepted by this cell and logged into the computer, together with basic information about the complainant, the subject of the complaint, the officer to whom it had been marked and the time frame in which resolution was expected. The complainant received an acknowledgement with a number on it, which itself was considered a novel step in those days.

We could now follow up with field officers, as we had a consolidated officer-wise list of pending grievances. The system of updating the applicant about the action taken through a letter was also initiated. This too was much appreciated and the development made it to Doordarshan, the singular television channel that devoted a special programme to it. Interestingly, the current Centralized Public Grievance Redress and Monitoring System (CPGRAMS) is quite similar to what we had done in Purnea in 1986!

I remember an interesting episode about computerization of public grievances. In 1988, Rajiv Gandhi, the then PM, had called the Collectors' Conference in Hyderabad. I mentioned the public grievances work to the director, National Institute of Rural Development (NIRD) and Panchayati Raj, J.M. Lyngdoh, an IAS officer from the Bihar cadre, who later became Chief Election Commissioner of India. He appreciated the efforts and desired that the software should be demonstrated before the PM.

However, there were no computers in the classrooms where the sessions were to be held. A computer was brought from the administrative section of NIRD into one of the classrooms. I installed on it my PG software, a non-Windows, non-graphic MS-DOS programme. I tested the software for demonstration and got ready for the PM's arrival. There was a moment of nervousness before I presented the various facets of the software to the PM.

The PM listened to the presentation—which was mainly about the purpose of each menu option, i.e., registration of complaint, its forwarding, reminder system and performance reports by the officer by subject or pendency—with great interest. He was curious and asked many questions. As a junior officer, I was happy that I could present my solution to the PM of the country.

Computerization of Treasury

In any district, the Treasury is one of the most important institutions. It keeps track of expenditure and income at the district level. Whatever money is spent from the consolidated fund of the state on anything, from employees' salary to the construction of a road, is withdrawn from the Treasury. Similarly, the revenue to be credited to the state is deposited in the Treasury.

The Treasury tracks expenditure in various sectors and helps in dynamic budgeting and decision-making. The consolidated head-wise statement of income and expenditure is expected to be submitted to the Accountant General (AG) and the state government at the end of each month.

Back then, the manual collation and consolidation of this information took minimum three to four months. There were cases of fraud too, where the money was withdrawn based on forged papers, without corresponding sanction or allocation. Incidentally, this modus operandi was similar to that of the Fodder Scam in Bihar that hit newspaper headlines a few years later, in 1996. Because of delay in processing the information, the system of keeping a running check on the availability of funds under each head or subhead was weak, resulting in a time lag in expenditure and income accounts.

As the system of Treasury was completely rule-based, it was a fit case for introduction of computerization. I coded the software for the Treasury and implemented it. Being a running system, it took me a few months to fully operationalize it. Also, there was no concept of connectivity, as all the machines were standalone.

Once the system became operational, we submitted the monthly accounts to the AG just after the conclusion of the month. Everyone was astounded. Some contended that the accounts may only be submitted on the prescribed, pre-printed forms, but later were persuaded that a computer printout that maintains the same structure ought to be equally valid.

The impact of a properly working Treasury is enormous and goes beyond the mere convenience of sending out quick reports. Doordarshan thought it was big news and dedicated a special report to it. The district received a lot of publicity for this.

I wrote detailed manuals for both these programmes, the Treasury and Public Grievances. Other applications that I wrote in Purnea included the monitoring of the National Rural Employment Programme (NREP) projects, and management information system (MIS) for the teachers and issues related to the disposal of their provident fund (PF).

Coding—The New Literacy: You could ask why I was coding the applications. Would it not have been better to hire a programmer for this work? At one stage, I asked myself the same question. After some search, we found a programmer in Calcutta (now Kolkata) and contracted him to write the software for handling public grievances for a lump sum amount of ₹50,000.

As Purnea has a sizeable Bengali population and is close to Siliguri, the programmer agreed to come and stay in Purnea for the duration of the software development. He was put up at the Circuit House. While he would write the code, I provided him the flowcharts with details of the procedure and other functional requirements. He worked for three months and produced about 400 pages of COBOL code. Unfortunately, he could not make the programme run! After multiple tries, both of us gave up and I concluded that it is much better to write the code rather than explain the requirements to somebody else and expect him to produce the software.

It was not easy to come across good software professionals in Bihar in those days. It was also difficult for me to personally explain the requirements in a systematic manner. Therefore, it appeared that it was better for an administrator to understand technology, rather than for a technologist to understand the administrative processes and rules, and then to write the code! This debate is still alive. A few years ago, Steve Jobs advised that everyone must learn to code. It triggered a movement in Silicon Valley which was duly countered with 'coding is not the new literacy'.[1]

I stayed in Purnea for three years, during which all the applications

---

[1] Available at: https://www.chris-granger.com/2015/01/26/coding-is-not-the-new-literacy/. Last accessed on 2 May 2020.

I wrote including for land-ceiling cases, NREP, the Treasury and Public Grievances had completely stabilized and continued to work even after I was transferred as Deputy Commissioner (DC) Dhanbad (in present-day Jharkhand state) in May 1989.

Assembly Elections, 1990

Different from Purnea's largely agrarian economy, Dhanbad is an industrial district with law and order issues related to the prominent coal industry. I was posted there for a little over a year and learnt about the coal mafia that operated in that area.

State Assembly Elections were held in Bihar in March 1990. It was known that criminals and history-sheeters were contesting these elections and family members of the coal mafia were in the fray. Booth-capturing was a serious problem, especially in our district. It was common knowledge that armed criminals were brought in busloads from eastern Uttar Pradesh (UP) to overpower the presiding officer and other personnel manning the polling booths. The votes could then be forcibly cast in favour of a particular candidate. However, the use of force was the second option. The first one was to befriend the polling staff and capture the booth with their assistance. If the first one did not work, the second option was always open to be exercised, which would scare away voters and the polling staff, leaving the goons to stamp the ballot papers in favour of their candidates. We did not have Electronic Voting Machines (EVMs) then.

I figured that if we delayed forming the polling parties till the last moment, nobody would know where they would be assigned the polling duty, i.e., to which booth or assembly constituency. This measure should preclude any pre-poll contacts and agreements.

A polling party usually consists of five persons under one presiding officer. These polling personnel are drawn from various departments of the state and central governments working in the district, including public-sector undertakings (PSUs). We had a few central PSUs in Dhanbad, the most important being Bharat Coking Coal Limited (BCCL).

We decided to 'randomize' the formation of polling parties. First, we collected the names of all the employees who were to be deployed for polling duties, with their levels and put these into data tables. Then, we wrote a programme to randomly fetch personnel from each table and form multiple groups (polling parties). I fully tested the programme before the polling date with the data. As it was random, every time we ran the algorithm, we got a different composition of parties! We got the order for the appointment of polling personnel served on all the personnel on poll duty, without giving them details of either the booth or the assembly constituency. Thus, while the employees knew that they will be sent on poll duty, they did not know where they would be deployed on the day of polling. There were more than a thousand booths in the eight assembly constituencies in Dhanbad, which required the services of around 5,000 polling personnel.

The final deployment letters were printed all night on dot matrix printers using pre-printed stationery. In the morning, we put up the names of all the personnel in alphabetical order. They were to check their names in the list to discover the assembly constituency allotted to them, proceed to the hall where they met other members of their party, collect the ballot papers and wait to be informed of the vehicle assigned to ferry them to the destination and back.

No Room for Mistakes: The conduct of the polls is a gigantic exercise that must be got right in one attempt. There is no room for mistakes. Although we had checked that the election law does not mandate the booth to be mentioned in the letter of appointment of the personnel, the new procedure was a gamble. If something went wrong, it would have had legal and constitutional implications. And the blame would have fallen on the DC, who was trying to experiment with an extremely sensitive and constitutional issue!

Perhaps, fortune does favour the brave. We were able to form the parties and despatch them in time without any hitch. The election was conducted peacefully. The maintenance of secrecy worked to our tremendous advantage in preventing prior 'alliances' forged with the knowledge of who was going where.

A downside of working in Bihar or indeed anywhere in eastern India is that your work is not recognized. The same process of randomization was adopted at the national level and heralded as an innovative idea much later, but nobody acknowledges that it was first employed at Dhanbad many years before it was conceived and implemented at the national level!

After the Assembly Elections, there was a new government in the state. I was transferred from Dhanbad and posted as State Transport Commissioner (STC) of Bihar, with headquarters at Patna, in May of 1990.

Streamlining Cumbersome Processes

The Transport department, responsible for the levy and collection of fees and taxes on vehicles, was infamous for being one of the most corrupt departments in the state. There was a system of middlemen who were an institution in themselves. Like lawyers have shanties in front of small tehsil offices, these middlemen had little huts in front of the District Transport Office (DTO). They offered services priced at varying rates for getting a driving licence, fitness certificate for transport vehicles, surrender of off-road vehicles for non-payment of taxes and re-registration of vehicles, which was typically done to change the identity of a vehicle to avoid payment of large tax arrears. There was also an Enforcement wing to enforce the provisions of the Motor Vehicles Act, 1988. Unfortunately, this wing worked more to harass drivers and extract money rather than enforce the provisions of the Act. Police officers and constables would make a beeline for a deputation to the Transport department, so that they could become a part of this extraction gang. They would bring the influence of the ministers or others for getting posted to the Transport department 'in the interest of state revenue.'

The Transport department, where processes are rule-based, was also perfect for computerization. I decided to begin with the Patna DTO. We started by digitizing motor vehicle records, including taxation history. First, we digitized the register for motor vehicles, series-wise, which were then printed and verified for their accuracy.

We replaced the old handwritten registers and started to put up tax-payment stickers on the relevant pages to keep the tax-payment history updated. We also printed a series-wise registration numbers of heavy vehicles that defaulted on tax and provided this list to the Enforcement officers to intercept these vehicles on the road. We also initiated enquiry into seemingly fraudulent re-registrations done earlier to avoid taxes and criminal liabilities. Extracting information became extremely easy, as vehicle details were digitized and simple database queries would extract the relevant information.

The process of depositing vehicle tax was cumbersome too. First, one had to go to the DTO to find out the demand; next, to the bank for a deposit challan in quadruplicate; then one had to provide the details of the tax proposed to be deposited. Finally, on depositing the money in the bank, you got two copies duly stamped, one of which was presented to the DTO office. The DTO, however, would also await the copy from the bank, which was matched with the one provided by the vehicle owner, to ensure that it was not forged! Only then the owner was issued the tax token for the vehicle. Obviously, this process had to be 'assisted' throughout, without which the tax token could never be issued.

People used to say that notwithstanding technology interventions, crooks would always find ways to make money, especially illegal ways. The Transport department was one of those places where you had to pay a bribe even to pay tax to the government!

Simplifying Tax Payments

We used information and communication technology (ICT) in several areas: registration, fitness certificates and taxation. However, the most important intervention was payment of the motor vehicle tax across the counter.

After digitizing the motor vehicle and taxation register, we wrote the software, with the assistance of the National Informatics Centre (NIC), to accept tax payment across the counter. This is how it worked: A person approaches the tax-payment counter and provides the vehicle number, which when entered into the computer, returns the tax due on the vehicle. The owner pays the tax in cash (or demand draft, if he has already made the draft after ascertaining the tax liability) and the tax-token is printed then and there. No more trips to the bank and the DTO's office.

There was no internet back then to facilitate online transactions as done today. We had to keep the servers in the offices and have multiple counters served by dumb terminals, for which we used Unix OS. This system was first introduced in the year 1992 in five districts: Patna, Muzaffarpur, Dhanbad, Jamshedpur and Ranchi. It was an instant hit. The revenue of the Transport department also went up three times—from a meagre ₹48 crore (₹480 million) to more than ₹150 crore (₹1,500 million) per year during the period 1990–94.

In May of 1994, I found myself 'shunted' to the post of Director of Treasuries and Provident Fund, which was considered an insignificant assignment, especially after being Transport Commissioner. The new position, however, did have significant opportunity for reforms, and my earlier experiment to computerize the state treasuries came in handy. The General Provident Fund (GPF) too needed urgent reforms. Prior to 1986, the job of maintaining GPF accounts of the state employees was done by the AG of Bihar. Thereafter, this was transferred to the state government, which had created district GPF offices and the Directorate of GPF at Patna.

Automating PF Accounts

The deduction of GPF was done at the time of payment of salary to the employee. A loose sheet of paper, called the GPF schedule, was attached with each salary slip and presented to the Treasury. After the salary was disbursed, the GPF schedule was sent to the GPF Directorate or the District GPF officer for updating the employee's ledger. These slips were sent in gunny bags and would fly loose if the bags got damaged, which was quite often, and people in offices walked over these torn bags and GPF schedules.

The information was almost never posted in the employee's ledger and therefore an updated annual statement, as required, was never sent to the subscriber. Getting a GPF advance was an uphill task because it first required the accounts to be updated and reconciled. Even if you never asked for an advance, you had to encash the GPF deposits upon retirement. Therefore, employees started keeping the certificates of deposits obtained from the salary slips and produced them as collateral evidence to bring the accounts up to date. The primary evidence of the GPF schedules was obviously non-traceable. These conditions were perfect for corruption to flourish and for the harassment of employees by the GPF offices.

We worked to computerize the GPF entries of officers of the All India Services (IAS, IPS [Indian Police Service] and IFS [Indian Forest Service]). This was relatively easy because the responsibility of maintaining the accounts was with the Directorate and there were only a few hundred such officers in the state. We sent them details of the deposits and withdrawals, whatever was available with us, and they responded with missing credits. Corrected and revised accounts were then sent to them for the first time in their careers!

Although, as Director of Treasuries and GPF for one year, I was unable to introduce any major changes, the knowledge of procedures came handy when years later, I worked as Principal Secretary in the Information Technology (IT) department, and we undertook the job of automating PF accounts. Understanding the system is the important prerequisite for rationalizing or automating government processes.

During this period in Patna, at the request of a friend who was a stockbroker at the Patna Stock Exchange, I wrote a software for stockbrokers, which was christened BD-FAST, an acronym for Bad Deliveries, Financial Accounting and Securities Transactions. He gave me the rules and I wrote the design and the code. We even sold a few copies of this software in Patna and Calcutta. I then started working on a new programming platform called PowerBuilder, which was a product of Sybase and included a database as part of its offering. However, this could not go far, as I was shifted to Delhi in the Department of Economic Affairs (DEA), Ministry of Finance (MoF) in the Government of India (GoI) in May 1995.

I worked in the DEA during 1995–2000, first as Director

and later as Joint Secretary. I dealt with international financial institutions such as the World Bank, the Asian Development Bank (ADB) and the International Monetary Fund (IMF) besides also working as Joint Secretary (Admin). The work was different from what it is in the state and I did not see much opportunity of using computers, except for organizing projects and their details in some database.

Keeping track of files and papers, however, was a nightmare. NIC had a solution, called Diary, which was installed on our computers. By now, we were equipped with a local area network (LAN) and a local server on which to store the data. Unfortunately, the NIC software allotted its own unique number to each file the first time its details were entered into the software. There was no way to search for a file unless you had this number handy and NIC was unwilling to support a search by the file's original number or subject.

So, I wrote a software called FileTracker on my own, using PowerBuilder, which I also showed to Naveen Kumar, then the joint secretary in-charge of administration, who later became the chairman of GST network. He was happy, and the DEA introduced FileTracker for use by everybody. It continued to be used for many years in that office.

Break from Serious Computing

As a full-time government officer in a senior position and a part-time amateur programmer, I had doubts about my own adequacy in the field of IT. I recalled a lecture on the Number Theory during my MSc days at IIT Kanpur, which was delivered by a senior IAS officer from the UP government.

He was introduced as someone with a deep interest in Number Theory. Unfortunately, his lecture was quite substandard. It was clear that he had lost touch with the field and his knowledge was dated. As students, we felt the lecture was a waste of our time, but the faculty and the head of the department remained respectful throughout, possibly due to his high position in the government. I wondered whether IT professionals were similarly deferential towards me and chose not to criticize me for the superficial knowledge in their field.

I decided to undergo a rigorous training programme in computer science and understand the subject before working on the issue of e-governance that interested me. Although the GoI sponsored mid-career study, it was only in areas such as public policy at Harvard or economics at the London School of Economics. Hardcore computer science wasn't in the bouquet.

Hence, I took the GRE and TOEFL exams, the pre-qualifying tests in international graduate-level coursework, and applied to several universities for admission in master's and PhD programmes in computer science. There were some offers of admission, including one from the Carnegie Mellon University (CMU) for a master's in Information Systems Management (MISM) programme, but without the offer of a scholarship. When I visited them, taking a detour from my regular visits to the World Bank headquarters in Washington, DC, they recommended I join the programme, as I would certainly get a highly paid job thereafter. But I had no intention of leaving my government job. So, I decided to study at the University of California, Riverside (UCR), which alone had made an offer of financial assistance. Though it was not substantial, I accepted it, as it was the best among other offers.

Struggle for Study Leave: All India Services Rules prescribe that an officer may avail two years' study leave, extendable to five years. I had two options: either to get my leave sanctioned by the GoI or to return to the parent cadre, i.e., Bihar, and get my leave sanctioned from there. The latter was attractive for the reason that the study-leave period would be counted towards the 'cooling-off' period in the state, thus providing an earlier opportunity to return to the Centre.

I went to Patna to explore this possibility and met Lalu Prasad Yadav, the de facto CM then, and pleaded for leave. When he jokingly asked if I wanted to switch to the corporate world after studying in a foreign university, I replied that no corporate entity will hire a bureaucrat and that too at my age! I was asked to meet various senior officers, the most important being

Mukund Prasad, Principal Secretary to the CM. I soon realized it would be nearly impossible to get leave from there. I may well have to go on an unauthorized leave.

Thereafter, I tried my luck with officials in the GoI. The then finance minister, Yashwant Sinha, supported my application for study leave. He had learnt of FileTracker and my interest in using computers for administrative tasks. The Personnel department too was encouraging as were others including my friend and batchmate R. Bhattacharya, Joint Secretary in the Cabinet Secretariat and Prabhat Kumar, Cabinet Secretary. An exception was made in my case. I was granted study leave on full salary at the end of my deputation to the Centre in a programme not sponsored by the GoI.

Testing Time in the US

I took a substantial advance from my PF, converted it into dollars and reached the United States (US) at the beginning of September 2000. My family followed a few days later. Half of this advance went into purchasing the air tickets! Shortly thereafter, when I joined UCR, I realized what I had got myself and my family into. The transition from a comfortable living in the diplomatic area of Chanakyapuri in New Delhi to a student's life in California was too much and too sudden.

Our tiny 700-square-feet house with a single bathroom was a godsend for the six of us, as it was on the UCR campus. The monthly rent for this accommodation was US$550, which was almost half of my total income of about US$1,100 per month. Getting a house on rent outside would have cost us more and become unaffordable.

We had no furniture, no bed and no transport. An Indian professor, Dr Satish Tripathi, and his family took pity on us and brought us a set of four chairs and a table. They also gifted us two cycles that their sons had outgrown.

On the academic front too, I was in for a jolt. I had always considered myself a bright student and a good programmer. However, this confidence soon gave way to depression and helplessness. Students in my class were half my age and their programming skills were way better. Their knowledge about computer science was much more current and extensive compared to mine. For them, it was just a continuation of their undergraduate degree. I was amazed at the speed of their fingers on the keyboard.

I would get stuck in home assignments. Admitting to young classmates, that I did not understand something was chastening at first. But, as I reasoned, if I did poorly, I would lose the teaching assistantship and must go back to India because studying on government salary alone was unworkable. So, I sought help and guidance in programming from my fellow students from India. Slowly, I too developed the required fluency.

Living in the US, I soon realized that we were used to a highly protected and privileged environment in India. For instance, I failed to get a driving licence twice before finally clearing the written exam and driving test in the third attempt. I had learnt driving early in my career in India. It was also ironical and insulting that a person who had headed the Motor Vehicles department in Bihar for four years had failed the drive test, not once but twice! Also, to ensure survival in the US, I was forced to increase my teaching-assistance time and had to take up some tuition jobs at US$20 per hour to supplement my meagre income. We really had to live a frugal life to subsist within a restrictive income. As a student, I had health insurance. However, I could not buy any insurance cover for my wife and children, as I had no money to pay for the insurance premium.

Eventually, two years passed and I got the master's degree in computer science with a GPA of 3.9 out of 4. It was a good score and my friends in Silicon Valley were surprised that I was headed back to India after completing the course. They advised that I spend at least one more year for Optional Practical Training (OPT) allowed to foreign students. They thought I could earn and test the waters in this period. Some even suggested setting up an IT company. When I showed disinclination, they offered employment at salaries upwards of US$200,000 per annum.

I told them that it had never been my intention to leave the job I had back home. I had come to the US only to study. Many thought it a foolish move to waste two years of study, if it wasn't going to bring any material rewards. Some thought that maybe it was pressure from my employer, the GoI, which had sponsored the study, but were plainly surprised that the government was least interested in what I did with my time in the US. I was used to my well-wishers thinking of me as idiosyncratic. They opined that my first foolishness was to waste two years doing computer science and the second one was to return to India.

After completing my studies at UCR, I returned to Jharkhand, my parent cadre created after the bifurcation of Bihar in November 2000, where I remained in different positions until 2009.

Using IT for Governance

Nobody in Jharkhand asked me what I had been doing in the US. No questions were asked about the university, subjects or the study programme. Largely, my colleagues and seniors thought that I had enjoyed a comfortable holiday in the US. I was posted as Principal Secretary to the governor of Jharkhand, Justice Rama Jois. We got along quite well.

After his transfer as Governor of Bihar, I got myself out of Raj Bhavan in June 2003 and was posted as Principal Secretary of the IT department of Jharkhand. The then CM, Arjun Munda, was quite interested in using IT for governance. For me, it was a posting of choice. After being transferred through nine posts in seven years between November 2002 and July 2009, I still put in four years as Secretary of the IT department, Jharkhand, a department nobody was interested in.

These years gave me the unique opportunity to initiate several e-governance projects in Jharkhand. We completed a state-wide area network project called Jharnet, way back in 2003–04. Other projects that I initiated included online registration (e-Nibandhan), commercial taxes, transport, municipalities, treasuries, GPF and salary payment. My software, FileTracker, was upgraded to a web-based system from the client server. It was an enjoyable time and we also received recognition and awards for our projects in IT and e-governance. Then, I got the opportunity to work on Aadhaar.

In June 2009, I received a call from the Cabinet Secretariat that I was being considered for a project of national importance and whether I would like to work with the GoI. My response was: 'Surely, but first tell me what the project is.' I was asked to meet Nandan Nilekani on my next visit to Delhi. There was an opportunity soon enough.

Joining a Start-Up

Even before the meeting, I had figured out its purpose. I knew that the Planning Commission had constituted UIDAI. Its mandate was to create unique IDs for the residents of India. Mr Nilekani had volunteered to take up the assignment and was nominated as the chairman of this body. Now, he was looking for a director general (DG) for the project. My name was suggested to Nandan by our mutual friend K.P. Krishnan (KP to all his friends), an IAS officer of the Karnataka cadre. This was going to be an exciting challenge. Technology fascinated me and also the idea that it could be leveraged to transform governance and public service delivery systems.

At the meeting, Nandan quickly came to the point. He was going to join as the chairman of UIDAI and was looking for somebody to work as the CEO. We discussed the project in general terms. When Nandan asked me whether I would like to 'do it', I unhesitatingly said 'yes'. When he asked me why, I responded that when a person like him was leaving the industry to join this project, it was motivation enough for me. At a minimum, it was just another 'posting' for me, but it was also a technology-driven project and I had enjoyed using technology for governance my entire career.

The meeting ended. I still remember Nandan's words: 'We have a deal then, OK?' I said: 'Yes, it is a deal.' This term 'deal' was not 'normal' for me, as in my bureaucratic lexicon, the word connotes an unholy nexus. Later, I learnt that some officers who were approached with the same offer had declined. This was also natural, as most bureaucrats want a job with a chair and chamber besides some minimum creature comforts. This was a job where we had to begin afresh.

The project was going to be like a start-up. This was uncharted territory, as the start-up would be within the ambit of the government. There could be special challenges in setting up the initial infrastructure and the organization.

Then came the usual politics of my indispensability from the Government of Jharkhand. The Cabinet Secretariat asked Nandan to suggest some other name, as I was not available. By that time, I had got so interested in the project that I briefed Nandan about ways to surmount these difficulties. Nandan moved the PMO to get me relieved from Jharkhand. M.N. Prasad, who had trained me when I joined the IAS and who was working as Secretary in the PMO, helped me get relieved from Jharkhand.

Nandan joined as Chairman, UIDAI, on 23 July 2009 with the rank of a minister of the Union Cabinet. I was present in his first press conference. When I joined a week later, on 31 July 2009, I had no office room. When I asked Nandan where I would sit (as there was no room for the DG), he asked me to add another chair in his office room in Yojana Bhawan, Delhi. It was a start-up, he would say. Of course, we got a room and an office a few days later. For the next three months, until October, it was only the two of us in UIDAI and we spent our time making presentations before various departments in Delhi and the state governments about the Unique ID project that we were going to do. There was a standard presentation and both of us were like salespersons! K. Ganga, an officer of the Indian Audit and Accounts Service (IAAS) became our Deputy Director General (DDG) Finance. She joined in October and others came in later.

Thus began my journey into Aadhaar.

Section I

THE WORLD'S LARGEST
IDENTITY INFRASTRUCTURE

Chapter 1

A UNIQUE PUBLIC-PRIVATE PARTNERSHIP

It can't be any new note. When you look at the
keyboard, all the notes are there already.

—Haruki Murakami, Japanese author

When I joined UIDAI, the design of its management structure was on my mind. Coincidentally, soon after UIDAI's ship took to the sea, Professor Hayagreeva Rao of Stanford Graduate School of Business wrote in the *Harvard Business Review*:

> When I ask MBA students and executives to design the job of a pirate ship captain, they invariably lump together two areas of responsibility: star tasks—strategic work such as target identification, command during battle and negotiating alliances to form fleets—and guardian tasks, which are operational work such as allocating arms, adjudicating conflict, punishing indiscipline, distributing loot and organizing care for the sick and injured.[1]

He said this was a mistake because candidates who can do both tasks exceptionally well are rare. The pirate ship, therefore, had a captain and a quartermaster general, each independent and non-interfering in the other's role.

'How do we divide our duties?' I asked Nandan. This question is normally never asked as the Rules of Executive Business clearly define the roles of the minister and the departmental secretary. However,

---

[1]Hayagreeva Rao, 'What 17th-Century Pirates Can Teach Us about Job Design,' *Harvard Business Review* 88.10 (2010).

this outfit did not fit clearly into a ministry or a department. There was nothing except the notification of 28 January 2009!

Nandan's response was clear: 'I will do the environmental management; evangelization and you work on implementation of the project.' He also assured me that he would provide me full independence in doing my job.

The UIDAI ship's captain, Nandan Nilekani, formally designated as Chairman, was the charismatic leader. I was the quartermaster general, designated as DG in the organization. The best voyages, I imagine, must involve a skilled team, a great ship and a treacherous route. The expedition that UIDAI undertook went into uncharted territory with unknown risks and severe weather conditions on the way. To survive, whether by design or luck, the team and the vessel had to be top-notch.

The Ship Assembles a Lean Crew

The notification constituting UIDAI described its composition, structure and mandate and stipulated a core team of 115 officials. The broad distribution was one DG in the rank of Additional Secretary, 36 DDGs at the Joint Secretary level; with one in the headquarters and 35 in the states and Union Territories. For complete staffing, as per usual norms of deputy secretaries, section officers, assistants, etc., the total sanctioned strength came to 1,331 personnel.

As the only two persons on the project, the first thing Nandan and I agreed upon was that we did not need so many posts. We could do with less and didn't need them in every state. One regional office to cater to a number of states or Union Territories could work well. Nandan wanted to reduce the sanctioned strength to about 200. I told him: 'We can, but you will not get a Param Vir Chakra for this act.' Having so many posts already sanctioned is a great asset because one struggles for years in the government to get the sanction for even a single post.

I reminded him that along the way, there would be parliamentary questions, references from VIPs, state governments, NGOs, civil society, privacy advocates and countless others. We would need staff, as UIDAI would be a pan-departmental organization and there would be paperwork and correspondence that would not contribute directly to our work, but would nevertheless need to be attended to. In the end, we decided to keep this number at about 400, but did not fill up all the posts as a strategy.

The unfilled posts in the book of sanctions could do no harm, but were available should the need arise. Thus, we were aggressive in action, but avoided cutting close to the bone. It's a theme that would recur many times over the years: be aggressive, but keep options for recovery from such aggression open, if it proved to be misguided.

As we were building the organization and assembling people, there were two main sources: private-sector employees and government employees on deputation. Most of these officers joined of their own volition, stepping out of the comfort of their services to work on the project.

A Diverse Public-Sector Team

Nandan wanted an eclectic team. So, we deliberately looked for talent in diverse services. This second point is crucial. Around the time when UIDAI was formed, the business and academic worlds were taken up by the idea that innovation happens at the intersection of diverse disciplines, cultures and industries.[2] Perhaps, sensing this *zeitgeist* and realizing that continual innovation was critical in driving implementation, Nandan had a vision of what the team needed to be. Let me introduce a sample of its senior members from the government.

Ganga was the third employee to join UIDAI, after Nandan and I. She joined in October 2009 as the DDG (Finance), or what in the corporate world is called the chief financial officer (CFO). But she wasn't your run-of-the-mill finance officer who examines

---

[2]Frans Johansson, *The Medici Effect: Breakthrough insights at the intersection of ideas, concepts, and cultures.* Boston, Mass: Harvard Business School Press, 2004. Print.

proposals sent for concurrence. 'Ganga Ma'am' was a gentle tigress who kept things in order, offered advice and fully shared in creative problem-solving.

Ashok Pal Singh joined UIDAI from the Indian Postal Service (IPoS). Nandan called him the 'strategic thinker.' He gave the idea that anybody with an Aadhaar number should automatically have a bank account and further ventured that Aadhaar itself could act as a financial address. Anil Kumar Khachi, an IAS officer of the Himachal Pradesh cadre, is methodical and liked to work silently. Eager to undertake a responsibility, he was appropriately put in charge of enrolments in Aadhaar. Dr Ashok Dalwai, another IAS officer of the Orissa (now Odisha) cadre, was Regional DDG in Bengaluru, who later also assumed responsibility as DDG for the technology centre. His job was tough. In a technology-driven project supporting continuous operations, ensuring security, development of technology and its deployment could be tricky with the overall processes at risk of being derailed due to bugs in the code. Ashok had grasped the essential architecture of the system and how it ought to operate. He effectively monitored operations and played a central role in the technology backbone of UIDAI.

Dr Ajay Bhushan Pandey, an IAS officer from the 1984 batch belonging to the Maharashtra cadre, was posted as Regional DDG in Mumbai. Besides ensuring onboarding of states such as Gujarat and Maharashtra falling under his region, he also worked as our in-house auditor of the technology centre and its processes. As an alumnus of IIT Kanpur with a PhD in computer science from the University of Minnesota, Ajay designed algorithms and sandboxes to check how we were doing on the issue of deduplication and other technology parameters. His passion and rigour immensely contributed to the working of our organization.

Sujata Chaturvedi and Keshni Arora, two outstanding IAS officers, worked in our team. Sujata worked as DDG of the Delhi region and took special interest in social-inclusion efforts of Aadhaar, organizing special camps to enroll socially marginalized and disadvantaged groups such as orphans, homeless and people afflicted with diseases such as leprosy. Other regions followed her example and efforts. Keshni, from the Haryana cadre, worked as Regional DDG, Chandigarh, and played a vital role in bringing the states—including Haryana, Punjab and Himachal Pradesh—in her region onboard. She also started some of the applications based on Aadhaar. Another key officer, B.B. Nanavati, from the Indian Revenue Service (IRS), was meticulous and took charge of procurement at UIDAI. After doing his homework, Nanavati would set conservative timelines that Nandan termed 'depressing'. But Nanavati always delivered on time!

I have mentioned these officers to illustrate the difference in their backgrounds, skills and experiences—the very difference that helped them make uniquely personal contributions to the success of the project. There were many more officers from services such as Indian Police Service (IPS), the Railways, Indian Telecom Service (ITS), Indian Defence Accounts Service (IDAS), etc., and of course, many others from outside the government too. They worked together as an inspired team. They were passionate, committed and driven to make the project a success. We shared successes and suffered setbacks together. It is, therefore, a little heartbreaking for me that I can't possibly name all of them here.

Private-Sector Recruitments

Our project had other peculiarities too. First, technology was going to undergird the project. Second, we were going to need significant resources in communication, demand generation, standardization, documentation, processes, enrolments and many more areas. As we were planning to do it in a way different from typical projects executed by government machinery, we decided to draw upon the expertise available in the private sector in these areas.

In the government, we normally work within hierarchical and well-defined reporting structures. However, since the last couple of decades, we take expert assistance from outside—either through consulting services or through the Project Monitoring Unit (PMU), a departmental body composed of outside resources. The significant difference at UIDAI was that we hired outside resources at top levels. For instance, the technology centre in Bengaluru had Srikanth Nadhamuni as the head of the team. He came from California's Silicon Valley and is a technologist and an entrepreneur. With vast experience in designing CPUs, developing software products and creating internet start-ups, Srikanth was extremely good with analytics and had a special talent for communicating technology ideas to bureaucrats like me. Being pragmatic and adaptable made him particularly suitable for this kind of a hybrid public project.

Similarly, in Dr Pramod Varma we had a chief architect who could innately grasp a potpourri of issues and find that elegant solution, which appears obvious only in hindsight. Even in Delhi, we put together a team of highly qualified resources to work alongside the government officers. Realizing that taking resources from consulting firms on secondment would not be cost-effective besides having to face issues related to reporting and ownership, we decided to engage the National Institute for Smart Government (NISG) as the expert agency to hire resources for us, dedicated to UIDAI on a full-time basis. These were highflyers from the private sector and academia, with the desire to work on a project of national importance.

We put together a vibrant team with both government and private resources working for the project. Sometimes, however, differences emerged in their selection, which had to be resolved. For instance, when we were recruiting resources for our PMU through NISG, Nandan had somebody in mind, whom I also knew. So, he asked me to engage the person. I informed Nandan that we had already advertised for the position and this person was one of the candidates in the ongoing selection process and we could have him onboard, once he emerged as the candidate of choice.

Nandan, however, had a different take. He was the chairman of UIDAI and the person ultimately responsible for the project. If he personally knew that a candidate is suitable, what else was required? Did he not have the final word as the chairman?

My response was that I had a completely rural and humble background. If knowing someone important had been the sole criterion, I could never have dreamt of landing my position at UIDAI. This project was being funded out of public resources, and hence, we had an obligation to provide equal opportunity to everybody based on merit. We could use the chairman's vote to break a tie in his favour, but nothing else. This was an emotional issue for me and one over which I was prepared to even leave the organization. Nandan understood and conceded the point gracefully. Decisions were based on discussion and reasoning and always had a 'logical basis'.

There were many occasions later on, at every level in the organization, when appeal to logic would help resolve differences rooted in the ethos of the organizations to which the individuals belonged. I think an open mind and the willingness to listen helped everyone learn, and prevented a fracture down the middle in the organization. Credit for developing such a culture largely goes to Nandan.

Forging an Ecosystem of Alliances

Presuming that recruitment was a formal, Union Public Service Commission (UPSC)-like process at UIDAI would be a mistake. UIDAI was not making entry-level recruitments nor was it necessarily for long-term employment.

We searched for candidates and invited them. We recommended who we knew, dipped into our personal network and kept an eye out for people we needed, even when we may not have known about that need. We also believed in giving equal opportunity to candidates who were brave enough to stand up and be counted.

For instance, at the annual InfoComm event, 2009, I had a chance meeting with Mohammed Asif Iqbal, a consultant working with PricewaterhouseCoopers (PwC). Asif had been blind since the age of 16 due to a genetic disorder. He graduated from St. Xavier's College, Kolkata, and earned a post graduate diploma in management (PGDM) followed by an MBA in human resources (HR) from the Symbiosis Centre for Management and Human Resource Development (SCMHRD)—both a first for someone totally blind.

Asif later joined UIDAI as a volunteer, on secondment from PwC, and helped make it easier for India's 70 million blind persons to be enrolled.

To enrol a billion residents and to provide them authentication services in which they could trust is not a matter of coffee, PowerPoint presentations, clever observations and a feel-good culture. It required careful thinking and formal procedures in mission mode. We couldn't have done everything on our own: those who tried that before had failed. Therefore, UIDAI had to forge alliances not only within the organization but in the ecosystem outside of it too.

At the heart of the ecosystem model was the willingness to trust other players in the system, to enable them to execute as a part of the whole. As my young colleague, D. Subhalakshmi, an HR professional who joined as a volunteer from the private sector, stated in an email to me:

> There was the culture of openness and the invitation for a diverse and heterogeneous group of people, agencies and organizations to contribute and participate. The volunteer or sabbatical programme invited private-sector members to join the core team in senior and important roles. There was a willingness to go all out and make the best use of all possible sources of support rather than to institute a control and command model, which would have inhibited participation from others.

We, thus, created a framework of volunteers, sabbatical and internship policies. Anyone with the right skills and the desire to work on the project was welcome. We accommodated the manner and time duration that enabled their participation. However, a job requirements description was especially created to weed out all but the best and the most determined.

When Nandan went out and made a public appeal to volunteer for the nation-building project, three senior tech team members were initially charged with handling queries and filtering out unqualified people. The team was looking for passionate, qualified, versatile and confident volunteers. The first-level requirements specified were: (i) you won't be paid, (ii) you won't have any title and you may not even get the authority to take decisions, (iii) there is no estimate on how long you can work here and (iv) your work won't be decided until you join.

As a result, we received responses only from those who were genuinely passionate, confident in their ability to make a difference and jack of many trades. These volunteer-applicants had the self-belief that they could thrive in an environment that did not even guarantee the usual benefits of service.

Nandan's charisma and the attractiveness of a project of national importance and scale brought together the best talent in the private sector, the academia and the government. For instance, we had Jagdish Babu from Intel, who came on a two-year sabbatical to manage the biometric device ecosystem. We had Professor Sanjay Sarma of the Massachusetts Institute of Technology (MIT), renowned for his work on radio-frequency identification (RFID), come onboard too. Other distinguished persons also joined. Raj Mashruwala, who was once Nandan's senior at IIT Bombay, heard of the project. He had moved to the US in the 1970s to study, and stayed there. He had set up a numbers system for the New York Stock Exchange (NYSE) and was doing well for himself. He came to attend a conference organized by UIDAI in July 2009 and was so fascinated by the diversity and energy of the group that he promptly moved to Bengaluru to assist in the project. He became the chief biometric coordinator.

UIDAI became a melting pot where dissimilar streams from the government, the private sector, civil society, academia and start-ups commingled. Naturally, with the variety of outlooks and approaches, there was bound to be some tumult and turmoil.

Uneasy Undercurrents

While largely UIDAI's objective became the common goal for all drivers in the project, Raj recounts many war stories that illustrate the emerging culture and the turbulence it created. For instance, in one of the first monthly status meeting, Viral Shah, a young expat from Silicon Valley and co-creator of Julia programming language, was the first person to arrive for the meeting. Due to his weak back, he sought a comfortable chair. He found one and sat down. As the support staff arrived, one could see in their faces that something was wrong. Finally, someone had the courage to tell Viral that he was sitting in a chair reserved for the chairman. Viral did not even understand the concept that certain chairs are reserved for certain persons. It is pretty common in an entrepreneurial tech start-up to intentionally maintain first come, first served model to emphasize equality.

In another instance, a volunteer began to work in the tech team while his official application was still being processed in Delhi. As a volunteer, he was not getting paid and not taking anyone else's job. He was in a rush to get going on his work because he didn't want to be a bottleneck for the rest of the team. When he changed his status on a social media site to indicate he was working at UIDAI even before he was officially accepted, he did not realize that he was violating a major code of conduct.

Within UIDAI too there were some clashes. When the teams started working together, there was huge excitement among both private and government personnel. Everyone had a sense of ownership of the project. Things went well for some time. However, there was a basic difference in the attitudes of private and government personnel. While those from the private sector were more adventurous and did not require 'entitlements' on day one, government 'deputationists' found themselves in an organization that did not have the 'entitlements' such as office rooms, peons, vehicles, etc., which are typically available in any government organization as soon as you join.

Additionally, in the government, there is much emphasis on processes and hierarchy, but not on presentation of thoughts. (Of course, the situation has changed since then, as communication is converted into a PowerPoint presentation in government too). Thus, while the government officers (GOs) are generally bright, they were typically not good at making presentations. On the other hand, the private-sector resources (PSRs) were more articulate and could make better presentations. In meetings, etc., the GOs felt that the PSRs were stealing the limelight, and they were being left out. The GOs also complained that they discussed ideas with the PSRs, who later, presented them as their own, thus unfairly taking the credit.

On the visibility and articulation issues, my advice to the GOs was: 'You simply cannot grumble about the visibility of private fellows. You must work hard, improve your presentation and articulation skills and then you will certainly have your say and become more visible.' This slowly brought some change as the GOs started to sharpen their presentation skills. But there were even larger issues that led to some disgruntlement.

Who Is the Boss?

Tensions started to brew largely due to factors of control and accountability. The GOs considered it to be a *sarkari* project and, therefore, one over which they should have full control in policy and decision-making. On the other hand, the PSRs also felt that they were hired to deliver and should have ownership and some decision-making powers. Typically, decisions in the government are taken on files based on notings or observations made by officials in the note-sheet or green sheet, as it is popularly called. As it was only the GOs who were making entries on these note-sheets, the PSRs started feeling left out of the formal decision-making process. This brought into sharp focus the issue as to who was the boss: the GOs or the PSRs?

The media too stepped in and fanned the flames. Somehow an impression was created in the media that the concept and technology for the project had come from the private sector. It was also publicized that the Indian diaspora of Silicon Valley had worked hard and actually delivered this project. I remember an article in *The Economist*[3] that stated that India, a poor country, did

[3] *The Economist*, 'Tribes Still Matter,' 22 January 2011. Available at: https://www.economist.com/special-report/2011/01/22/tribes-still-matter. Last accessed on 3 May 2020.

not know how to create identity until the Indian diaspora came to help Nandan.

Sadly, some of the people who had worked as volunteers with the project were feeding these stories that were a constant irritant to GOs working in the project. Unfortunately, the GOs had no means to contradict these reports due to the constraint that they could not go to the press.

Even those from the PMU gave interviews on television channels about the project. The general refrain from the GOs was: 'If something goes wrong, we will be held responsible, but real decisions are being made by the PMU and the Technology Development Unit (TDU). This is like having power without accountability, which must change. The PMU should only provide us inputs based on their area of expertise and work assigned to them.' But the situation worsened.

As we had introduced a culture of open communication in the organization through emails, most personnel, especially the PSRs, marked copies of mails to many others, crossing hierarchies. Discussions on emails also started to be treated like decisions and agreements. While the PSRs had no problems with the emails, the GOs felt that this would constitute a violation of proper procedures and there will be no official documents to trace the decision-making process, should there be any enquiry in future.

The GOs felt that this was a government project subject to all the rules of the government, which meant that should something go wrong, it is they (the GOs) who will be held accountable, and not the PSRs. Hence, the status of the PSRs could at best be advisors or consultants, but not the owners.

As I was a *sarkari* fellow and Nandan came from the private sector, a kind of vertical divide started to build up and the *sarkari* crowd started openly complaining to me. These fissures disturbed me. Nandan had also expressed his disappointment at these fissures at various points of time. Good thing was, we never took sides. The situation would have gone really bad if we had behaved like leaders of our respective groups—Nandan of the PSRs and I of the GOs.

However, on the issue of ownership, we were clear that it was a *sarkari* project and we would have to follow all the rules of the government and it is the GOs who would have to take the ownership and responsibility. Thus, while the PMU had an important role to play, the decision-making process had to be driven by the *sarkari* people. But this was not happening and there were some uneasy undercurrents.

I was worried that those in other departments had started to look at this project as a private project driven by Nandan. To make matters worse, in April 2011, a report in the magazine *IEEE Spectrum*[4] also emphasized the theme of Indian expats driving this project, stating that as efficiency was not a strength with most government bureaucracies, Nandan had turned to Silicon Valley and a core group had worked as unpaid volunteers until Nandan's office prepared the paperwork to give the team official authority.

This riled me and I finally wrote to Nandan drawing his attention to other similar stories in the press, while stating that these articles made us look like a bunch of idiots who had no idea about the project until Silicon Valley showed us the way. It was an insult to our country.

Diffusing Discord

I thought that Nandan would try to defend these guys, but he agreed that expats from Silicon Valley had 'no business misrepresenting and magnifying their contribution to the UIDAI project'. In a mail sent to me, he wrote: 'I think we have a larger problem, which is that my original goal of integrating the best knowledge and skills from outside into the government on a sustainable basis has failed completely. We should discuss that sometime.'

Nandan had acknowledged the problem and pointed to the real goal we had set for ourselves. We were together on the dire need to integrate outside expertise and skill into the government. I shared my own viewpoint that 'the divergence in the incentives,

[4]Joshua J. Romero, 'Fast Start for World's Biggest Biometrics ID Project,' *IEEE Spectrum*, 25 April 2011. Available at: http://spectrum.ieee.org/computing/it/fast-start-for-worlds-biggest-biometrics-id-project. Last accessed on 3 May 2020.

perception, approach, work culture and motivation between the people from the government and outside were some of the reasons for this non-integration. We will have to come up with solutions to reduce some of these divergence points'.

I think the larger lesson learnt was that we should clearly articulate the roles of people, including the hierarchy for any government organization in India for it to work efficiently. This was one area, I admit, where we could not fully harmonize the private and public teams. The conditioning due to the environment probably could not be changed over a short period. However, the tensions that had initially threatened to derail the project were gradually diffused and over time, the organization accepted this reality of coexistence of private- and public-sector people. In fact, working together considerably improved the understanding among team members and we were largely able to have a good working environment.

Trust in other people and respect for another's point of view worked to our advantage. However, looking back, it appears that there was something even more fundamental than trust at play here. It bordered on faith. We seemed to have tapped into the collective confidence of every individual working on the project, within or outside the organization, to solve the problems in their respective areas. We often forget this lesson in the rough and tumble of everyday life. But when it is a matter of life and death, in the trenches, nothing else seems more important.

UIDAI as an organization could not have thrived, except as a learning organization. Everyone was there, from the government or otherwise, because they had volunteered in some sense. This self-selection bias was reflected in the people's attitude to learning and the desire for new experiences. It was the power of ideas and the innate curiosity the team brought to the project that was a critical driver of its success.

These people had an appetite for learning. The eclectic mix of speakers we got through the intramural series that Ganga organized with people such as Sir Tim Berners Lee, inventor of the World Wide Web, Pranav Mistry, computer scientist and economists such as

Esther Duflo and Montek Singh Ahluwalia or even Abhijit Banerjee from MIT and Rama Bijapurkar, market strategist. Each became a source of inspiration.

Members from different services in the government gave up their hierarchies and divisions and learnt to work together. We created a structure in which one DDG and one senior private-sector member worked on key areas together, thus ensuring that both the worlds were united even at that level.

There was learning even in adopting the government's procedure of writing notings on files. As Subhalakshmi wrote:

I love the rigour and structure that file notes require—in evaluating options, clearly articulating the pros and cons of each, the robustness of the assessment and documenting all of this so that it can hold up to scrutiny. Clearly, the highly demanding nature of the reviewers added to the sharpness of the thought process and pre-work that needed to go into it, so that it was cleared in one go.

While giving everyone the autonomy to succeed, we used technology to isolate the impact of any mistakes. Without the capabilities of the diverse group that was created, the project could have hardly succeeded. But a great team, like the one at UIDAI in its early days, is more than the sum of its members. Its full potential comes from individuals supporting and amplifying the capabilities of each other. The UIDAI leadership fostered a culture in which everyone's contribution was valued, from the simple to the fundamental, so long as it helped in the realization of the ultimate goal.

There are many other lessons to be learnt from this unique experience of a private-public partnership. While creating an ecosystem that combines the best of both worlds may be easy, making it work does offer significant challenges. If one can handle these challenges with understanding and sensitivity without sacrificing the rigour and accountability principles, it can deliver the desired results.

Chapter 2

A RANDOM NUMBER BY CHOICE

Make it simple, but significant.

—Don Draper in *Mad Men*[5]

India issued the first Aadhaar number on 29 September 2010 to Ranjana Sonawane in Tembhli village of Maharashtra, in the presence of Prime Minister Dr Manmohan Singh and Congress President Mrs Sonia Gandhi. Today, Aadhaar has achieved almost universal coverage with more than 1.3 billion numbers issued. Due to the uniqueness of the number and the availability of online authentication services, Aadhaar has become central to India's public service delivery reforms.

It is true that many aspects of the Unique Identification (UID) Number are at variance with typical identity systems existing either in India or elsewhere in the world. It has introduced some new paradigms, breaking long-established stereotypes about ID systems. It is partly due to a lack of understanding of these facets that Aadhaar's design and operational aspects have been criticized.

Right from the beginning, many have questioned the basic design principles. As implementation progressed, many criticisms were muted. Nevertheless, there remain aspects that are not fully appreciated, some controversial or simply interesting.

In its Strategy Paper in August 2009,[6] UIDAI made an important but counter-intuitive decision that it would issue only numbers and

[5]Donald Francis 'Don' Draper (portrayed by Jon Hamm) is a fictional character and the protagonist of the AMC television series *Mad Men*.
[6]'UIDAI Strategy Overview: Creating a Unique Identity Number for Every Resident in India,' UIDAI, Planning Commission, Govt. of India, April 2010. Available at: https://www.prsindia.org/sites/default/files/bill_files/UIDAI_STRATEGY_OVERVIEW.pdf. Last accessed on 5 May 2020.

not cards. Many were surprised. Identities are always issued in the form of ID cards, isn't it? People had never heard of an identity number. What does it even mean? Even those who understood the concept of an online identity system criticized the decision, claiming that a card is the more appropriate form for India because poor connectivity would render an online system inoperable.

When people asked me to explain the online ID, I used to provide them an analogy of body (*sharir*) and soul (*atman*). The Aadhaar number is like the soul, indestructible and permanent (*ajar, amar*), as the Gita describes it. The paper or the card is its temporary abode (*sharir*), which holds the number (*atman*). When you have direct access to the *atman*, the *sharir* and its perishable nature are unimportant.

Just a Number, Not a Card

The decision to make Aadhaar just a number has greatly contributed to the goals of cost reduction, determination of uniqueness, inclusion of the poor and in enabling Aadhaar to become a digital ID platform, rather than a standalone smartcard. There are multiple reasons for the success of this design.

Lowest-Cost Solution: Aadhaar has not only proved to be robust with unique features otherwise not found in other ID systems, it is also the most frugal of solutions. For instance, a smartcard would have cost ₹70–80 at the very least. Add to this, the cost of couriering the card and the personal identification number (PIN) separately and securely. At ₹100 per smartcard, the total expenditure just for the smartcard would have been ₹12,000 crore (₹120 billion) for the 1.2 billion registrations. That is more than the total expenditure UIDAI has incurred till date. The recurring costs and inconveniences for a smartcard-based solution would have been of a much higher magnitude to all the stakeholders such as UIDAI, residents and corporates using the system.

Determination of Uniqueness: Identity in a centralized repository can be managed to weed out duplicates, should any be detected, because cancelling the duplicate Aadhaar is as simple as blocking the use of a number. Not so with ID systems that depend upon a physical token. If sufficient by itself, such a token continues to work as the ID till it is retrieved or destroyed. If you can't retrieve a duplicate, you cannot extinguish its use with any degree of certainty.

Can you prevent the use of a few currency notes—by serial number—out of a billion others? Not if people accept currency notes at sight. In fact, it is difficult to prevent the use of counterfeit currency despite it being printed on security paper with markings that facilitate identification. It is the same with other identities not linked to a database against which their use is verified through an online process every time.

Safeguarding Inclusion of Poor: Those who lose a smartcard or any ID token would also lose their identity and suffer enormous harassment to get it back. Further, the impact of such a loss would be especially severe among the poorest (who could lose even their homes to floods), because to access their entitlements, they would first need to prove their identity. The Aadhaar system is convenient for residents, absolving them of the worry of loss or misuse of their card. Today, millions of Aadhaar letters are downloaded and printed every day. Imagine the trouble and expenditure in reissue, if Aadhaar was a smartcard.

Digital ID Platform: A unique number that can be plugged into databases of names of users or beneficiaries of other systems serves the larger purpose of cleaning those databases while authenticating identities. Similarly, it also contributes to the goals of speed and scale besides future-proofing of technology, while keeping the cost of building such third-party systems very low. It is also much safer in terms of data security because the biometric data is not stored in billions of cards, but rather safely with UIDAI.

Smarter Than a Smartcard: Aadhaar as an online digital ID is different from identity tokens. Aadhaar is a digital ID infrastructure, while a smartcard (or any other card) is just an offline token. And tokens have lost all relevance in a digital and connected world. This phenomenon is plainly visible in many areas. For instance, earlier, airline tickets used to be coupons in an attractive jacket, which were exchanged at the check-in counter for the boarding pass. Today, however, you get a message on your mobile phone, an SMS or an email, which you may print, but don't need to. What's vital in the message, among many other information, is the passenger name record (PNR)—again just a number. There is no need for a ticket or a smartcard or any token because you have continual access to all the information related to your booking, via this ID called PNR, at the back end. Similarly, you can authenticate your identity online with just the Aadhaar number.

The Aadhaar number is communicated in a letter to the resident, with a perforated portion that bears her name, photograph, date of birth, gender, address and, of course, the Aadhaar number itself. This 'card' portion is about the size of a business card and only as durable. In retrospect, we may have overplayed the 'Aadhaar is a number and not a card' bit, because the Finance Ministry would not allow us to issue the number written on a piece of plastic as that somehow made it a card. Of course, a ₹5 plastic card is entirely different from a smartcard with an embedded chip, which would have cost about ₹100.

To this day, the letters that UIDAI issues are on paper. The residents do get it laminated and may get it printed on a plastic card, if they wish. But whether written in sand, papyrus or tattooed on the body, the Aadhaar number remains just as effective.

Smartcards, on the other hand, have several lifecycle management issues and are getting obsolete. They require PINs to be conveyed to the holder separately and securely. Update of demographic or biometric data requires collecting the old cards and issuing new ones. Cards can be lost, which would require the entire process to be repeated. Smartcards also need a smartcard reader. These issues create a logistical nightmare when more than a billion cards have to be issued in a country like India, where the address for some residents, could be 'behind the temple' or 'below the flyover.'

If Aadhaar was issued as a card, it would have become another ID card, just like a driving licence, a ration card or a voter ID card, which are essentially eligibility cards, authorizing one to drive, get rations from the public distribution system (PDS) or vote, respectively. Aadhaar has no eligibility attached with it. Hence by itself, an Aadhaar card would not have any value except probably enabling the holder to enter airports or to board trains or to check into a hotel. Those things are already enabled by existing ID cards. Aadhaar was meant to be an ID that could be combined with any transaction and work as proof of ID for accessing any formal system, such as PDS, Permanent Account Number (PAN) cards, Electors Photo Identity Card (EPIC) cards, Mahatma Gandhi National Rural Employment Guarantee Scheme (MGNREGA) job cards, or health insurance card, pensions and other databases. This mechanism could then also work as a cleansing agent to eliminate duplicates and fakes from these databases.

Now that Aadhaar is being used for various types of transactions, people have come to realize that the system is working as a trusted third-party authenticator of identities in digital transactions, and its utility as a number, and not a card, is being appreciated. Thus, Aadhaar is a token-less digital identity number that can be embedded into any database. With identity data residing at the UIDAI back end, it is possible for Aadhaar to work as a trusted and common authentication system for a person's identity in systems such as banking, provide proof of presence, lifting of ration from PDS shops and get mobile SIMs, among others.

By taking the decision to develop an online authentication system, we annoyed the industry that supplied smartcards for credit cards, metro systems and other uses. Unfortunately for them, in place of the shiny, physical and 'modern technology' that the smartcard represented, we offered a 'dumb' 12-digit number—printed on plain paper!

I hope the reader will agree with me that the choice of making Aadhaar as only a number and not a card has significantly contributed to the achievement of many overarching goals of the UID system.

It's a Random Number

Why is the Aadhaar number a random sequence of 12 digits? Why should its structure not provide some intelligence? It could have had codes for states, districts, gender, etc., embedded into the number itself. For example, UP09/12345 could be a number for a person from Agra district of UP.

As each alpha character can accommodate 26 values, against 10 for numerical digits, an alphanumeric system could make the Aadhaar much shorter in length. However, it was not considered a good choice in a multilingual society like ours with a high level of functional illiteracy. There are other reasons why Aadhaar is numerical digits only, such as: a digit is something that everybody understands; Arabic numerals are already there on the mobile phone and all other keypads; digit-only numbers, such as phone numbers, are easy to remember and can even be transmitted from the dial pad of TouchTone instruments.

The random UID numbering scheme was designed after a careful consideration of all the options and trade-offs. An internal report titled 'A UID Numbering Scheme',[7] became the basis of designing the structure of Aadhaar number. Some of the findings of this report are summarized below.

The number should also be large enough to accommodate future requirements.[8] In the case of Aadhaar, the 12 digits, with the 12th digit being a check digit[9], accommodates up to 100 billion numbers! Even the few restrictions relating to the use of the first digit, leaves us with 80 billion assignable numbers.

Contrarily, in an unconnected world, embedding intelligence

---

[7]Hemant Kanakia, Srikant Nadhamuni and Sanjay Sarma, 'A UID Numbering Scheme,' May 2010. Available at: https://archive.org/details/A_UID_Numbering_Scheme/mode/2up. Last accessed on 5 May 2020.

[8]Internet address schemes in version 4 and Y2K are some examples where the namespace ran out, requiring upgrade at a substantial cost. Imposing a structure on the format also results in unnecessary waste of space, such as the forced allocation of class A address blocks with over 16 million IPv4 addresses because the next lower type (class B) with 65,536 addresses was insufficient for some entities.

[9]Check Digit: Available at https://en.wikipedia.org/wiki/Check_digit. Last accessed on 11 May 2020.

in a number was useful, and hence adopted, because you could get information about the number 'at sight'. The postal code, for instance, indicates a geographical area without a lookup. For example, 1100xx indicates a postal index number in Delhi area. The credit card number reveals the payment corporation that issued it, such as Mastercard or Visa.

Since the beginning, Aadhaar, on the other hand, is designed as an online digital ID, with real-time access to the information associated with the number. Hence, there is no need to embed intelligence in the number. It has been designed not to disclose any information, personal or otherwise. Embedding intelligence violates this principle. The US's Social Security Number (SSN), for example, suffers from these problems because it is possible to guess information about a person from his SSN. Similarly, the patterns in the SSN enable a person to guess the number from the personal details of an individual. Hence, the US's Social Security Administration discourages the use of the SSN as a personal identifier.

As UID is a random number, it is impossible to guess any part of it from details you may know about the person, and equally the number itself does not disclose any information about its holder. A semantics-free 12-digit number was chosen to be Aadhaar, where the last digit is the check digit—to minimize data-entry errors—and the first digit is reserved for denoting the number type.[10] Rest of the digits are generated in a random manner. Only about 1 per cent of the numbers are actually assigned from the available universe of numbers (100 billion). Therefore, the probability that an Aadhaar-style number that meets the format and checksum requirements, is actually assigned to someone is pretty small. Therefore, to guess the Aadhaar ID of a person is also difficult, as every 12-digit number is not an Aadhaar number. To guess someone's Aadhaar number, working day and night for 1,500 years, would still give you less than a 50 per cent chance of success. Even

---

[10]The first digit is reserved as '1' for entity-related UIDs and the number is extensible with more digits, should such a need arise in future.

then, you would only have a number that you cannot directly use to authenticate the person.

Aadhaar is for lifetime. Once a number is assigned to a person, it remains with him or her even after death, and will not be reassigned. This avoids confusion regarding the ownership of the number at different times. You can literally say the the only thing permanent for a person is Aadhaar!

Not Limited to Citizens

Since the beginning of the project, a debate has been: why should Aadhaar be given to all 'eligible' residents, and not be limited to citizens only?

Actually, the decision to provide ID numbers to residents was not taken by UIDAI, but by the GoI. The mandate of UIDAI, as stipulated in the notification of its establishment itself, is to provide a unique identity number to the 'Residents of India'. 'Resident' means a person usually residing in India as defined under the various laws of our nation. The criticism is that UIDAI registers and enrols indiscriminately, without first determining the legal status of an individual.

Could issuing Aadhaar have been restricted to citizens only? What would that have entailed? Was there a downside to such a decision? What we do know from previous experience and what we can anticipate would elucidate these issues.

Going back in history, in 2003, the government had undertaken the exercise of providing Multipurpose National Identity Cards (MNIC) as a pilot project in a few subdistricts of selected districts in 13 states and Union Territories.[11] This was done on the recommendation of the Kargil Review Committee, constituted by the GoI on 29 July 1999, which submitted its report in January 2000. The experience was that determination of citizenship through such an exercise at a massive scale was fraught with serious problems.

---

[11]Taha Mehmood, 'India's New ID Card,' in Colin J. Bennett and David Lyon (ed.), *Playing the Identity Card: Surveillance, security and identification in global perspective.* London New York: Routledge, 2008. Print.

There were chances of illegal immigrants acquiring MNICs, as they possessed all kinds of papers such as ration cards, EPIC cards, etc., to buttress their claims regarding citizenship. Hence, the Ministry of Home Affairs (MHA) concluded that rather than preparing the National Register of Indian Citizens (NRIC) (as provided under The Citizenship Act, 1955), they should first prepare a register of Indian residents. The National Population Register (NPR) is a register of Indian residents, and not a register of Indian citizens. Rules under the Citizenship Act were subsequently amended to introduce the concept of NPR. Thus, the MHA is also enumerating residents and not restricting its enumeration to only citizens in the NPR exercise. If the MHA, which deals with Citizenship and Security, has found it difficult to prepare the register of Indian citizens, how could one expect UIDAI to issue ID numbers only to Indian citizens?

Even if one presumes that there was some way for UIDAI to determine the citizenship of an individual and they issued Aadhaar number only to such persons, then the fact of possessing the Aadhaar number would have become proof, recognized by the state, of one's citizenship. This means an error in determining one's citizenship status would have created a problem rather than solving one, besides the fact that it would have been a non-scalable exercise because determination of citizenship is a time-consuming task in itself. The fact that Aadhaar is not proof of citizenship has been incorporated in the Aadhaar Act, 2016,[12] and there is an explicit declaration on each Aadhaar letter that it is a proof of identity, and not of citizenship.

It is also important to note that the evolution of unique IDs started in 2006 with the exercise of providing unique IDs to below the poverty line (BPL) families by the Department of Information Technology. The focus was obviously on service delivery and development and not security or citizenship. Hence, the main purpose and direction of UIDs till date continues to be service delivery and development, and not determination of citizenship.

---

[12]Section 9 of the Aadhaar Act, 2016.

Designed to Establish Identity

One of the earliest suggestions was to collect all relevant details of a resident, including driving licence number, PAN card number, EPIC number, BPL card number, ration card number, etc., at the time of enrolment itself so that these numbers could be linked then and there. However, UIDAI's only function was to uniquely establish identity, or as Nandan used to say, that person X is X. Nothing more and nothing less. This ensured that the information collected could not be used for any other purpose, such as profiling.[13] This minimalistic approach is one of the features of Privacy by Design (PbD),[14] which was, in fact, incorporated in the Aadhaar Act, wherein UIDAI was specifically barred from collecting information about race, religion, caste, tribe, ethnicity, language, records of entitlement, income or medical history.[15]

UIDAI had no objection if extra information was collected by the Registrars—typically state governments—for their own purpose and use. This extra information was called Know Your Resident Plus or KYR+, while the basic information for Aadhaar enrolment was called Know Your Resident or KYR. Some state governments did notify collection of other attributes under KYR+. However, the results were not encouraging. Residents mostly came with documents sufficient only to prove their identity and address for their enrolments. Effectively, very little KYR+ data was collected, as enrolment agencies did not have any incentive to do additional data entry or to send back the resident if he or she did not bring KYR+ documents. They took the convenient line that the residents did not bring the extra papers, but they could not refuse Aadhaar enrolments to them.

---

[13]The Rwandan genocide of 1994 was much worse because their ID papers listed ethnicity. Following the death of the Rwandan president, a Hutu, about 800,000 Rwandans were killed in 100 days. Most of the dead were Tutsis, and most of those who perpetrated the violence were Hutus.

[14]Privacy by design (PbD) is an approach to systems engineering that seeks to ensure protection for the privacy of individuals in the development of products, services, business practices and physical infrastructures.

[15]Section 2(k) of the Aadhaar Act, 2016.

Of what use is Aadhaar then, without entitlements or benefits? A basic tenet that UIDAI put out in its Strategy Document was: UIDAI will only provide a unique identity number and this by itself will not ensure any entitlements, rights or guarantees. Aadhaar is deliberately and carefully designed to preclude any other use. Why then should people enrol for it?

An eligibility document names an eligible person who may, therefore, use the document as a de facto ID proof. A voter ID card, a driving licence or a ration card could all serve as identity documents in different situations. However, there are problems associated with such use of these documents, including no guarantee of their uniqueness or uniformity of information. For example, there are instances of people having multiple PAN cards and ration cards or even a different age in different documents.

Changing personal attributes in different cards may be advantageous to avail benefits from as many domains as possible— you become a senior citizen for railway travel and a young man for insurance purposes. Further, these documents can be easily faked, as there is no way to verify their authenticity without following a time-consuming procedure. Finally, these are not universal. For example, a voter ID card will only be issued to those who are 18 years of age or above, while ration cards are family documents as opposed to individual IDs.

Hence, unique ID was designed as the foundational ID document which is sufficient to prove the ID of a person and which, in turn, can be used by all the domains that issue eligibility documents. In other words, it can work as an ID platform on which eligibility applications can be built, including citizenship. Its availability on the digital platform makes it amenable and pluggable for use by various domains.

Authentication and eKYC

People have started to appreciate the usefulness of authentication in banking transactions, in PDS delivery systems and more trivially, in biometric attendance systems. The purpose of authentication is to enable residents to prove their identity and for service providers to confirm that residents are 'who they say they are' in order to deliver services and give access to benefits.

The eKYC is slightly different from simple authentication. It avoids the need for the resident to provide several pieces of information, each to be individually authenticated by UIDAI. It also obviates the need to provide a picture to the agency seeking the resident's identity information, by sharing the one that UIDAI already has. In other words, eKYC is the process by which UIDAI issues the agency, like a bank or SIM-card provider, a digitally signed electronic ID document after biometric authentication and authorization of the individual.

As UIDAI maintains only identity information about the residents, this information alone does not do much. However, it is important for every service delivery organization to establish the identity of its customers before delivering the services. It ensures that the service intended to be provided to A is provided only to A and not to B. This work of establishing A's identity is done by UIDAI. Hence, at the time of service delivery, UIDAI becomes the trusted third party to authenticate identities. This service can be combined in any digital transaction besides being used in a physical transaction such as allowing access to someone following provision of ID proof. However, it is vital in digital transactions, as it provides traceability and transparency of the transaction.

Conceived as the next-generation online ID system, Aadhaar's unconventional choices in design and execution were not fully understood, leading to avoidable controversies. However, as India marches deeper into the digital world, the merit of these choices is becoming apparent.

Chapter 3

THE FALLACY OF
TECHNOLOGICAL IMPOSSIBILITY

*I have not the smallest molecule of faith in aerial navigation other
than ballooning, or expectation of good results from any of the trials
we hear of. So you will understand that I would not care to be a
member of the aeronautical society.*

—Sir William Thomson (also known as Lord Kelvin)[16]

UIDAI isn't the Unique Authority of IDs in India, though it may
have earned the sobriquet for itself too. 'Unique' in its name
denotes the task that was set for the Authority: to assign one and
only one unique ID to each resident of India. From technology
experts to sociologists, many declared this to be an impossible task.
They pointed out, and rightly so, that nobody had accomplished it
in the world. Hence, how can you? That was a useful starting point
because we could start with a catalogue of the failed approaches to
avoid. The failure of obvious solutions also indicated that without
technological innovation, the goal would remain elusive.

More than a billion enrolled individuals and a million
authentications every hour on an average day should be enough
evidence to accept that Aadhaar is technologically sound. But this
is from the proverbial 20/20 hindsight. The solution was much less
certain when UIDAI started to work on it in 2009.

Looking back to the roots of Aadhaar, the Department of
Information Technology, Ministry of Communications and
Information Technology was given the job of determining how

---

[16]Lord Kelvin, Irish physicist, engineer and president of the Royal Society of England,
in a letter dated 8 December 1896, written to Lord Robert Baden-Powell, founder of
the worldwide scouting movement.

the government could assign unique IDs to BPL families to better manage the benefits and subsidies to them. This project had come out of the recommendations of the Empowered Group of Ministers (EGoM) constituted in 2006 under the chairmanship of the then minister of external affairs, Pranab Mukherjee. It was approved on 3 March 2006 at an estimated cost of ₹46.70 crore (₹467 million).[17]

Thereafter, a committee to suggest the processes of updation, modification, addition and deletion of data and fields from the core data was constituted under the chairmanship of Dr Arvind Virmani, principal advisor in the Planning Commission. The initial approach was to 'enable a seamless process of integration, updation and validation of information in a manner that would ultimately ensure realization of a central database which is exhaustive, accurate and complete and shared across all government departments.'[18]

One idea initially tried was to utilize existing databases of people relating to voter IDs, PDS, BPL families, etc. The idea was to test the feasibility of creating a database by mixing and matching the existing databases. Wipro was appointed as a consultant to this committee to examine the feasibility of this exercise, among other things. The consultants produced reports in multiple volumes, but concluded that this would be a futile exercise. For one, there was such non-uniformity in the way names and addresses were written in different databases that the same individual could not be identified and related in both.

Clearing the Deck for UIDAI

The Registrar General of India (RGI) was also engaged in the creation of NPR and issuance of MNICs to citizens of India. With the approval of the PM, it was decided to constitute an EGoM to collate the two schemes—NPR under the Citizenship Act, 1955, and the UID scheme. The EGoM was also empowered to look into the methodology and specific milestones for early and effective

---

[17]"Administrative Approval for the Project "Unique ID for BPL Families" Dated 3 March 2006.'

[18]Minutes of the third meeting of Process Committee on Unique ID for BPL families

completion of the scheme and take a final view on these. The EGoM was constituted on 4 December 2006 and a series of meetings took place during the period. In its first meeting on 22 November 2007, the EGoM affirmed the need for an identity-related resident database, whether based on *de novo* collection of data or on the basis of existing databases, such as the voter list. It also recognized the need to identify and establish an institutional mechanism that will own and maintain such a database.

The EGoM met again thrice in 2008 (on 28 January, 7 August and finally on 4 November) and decided to constitute and notify UIDAI as an executing authority. Decision on investing it with statutory authority was left for the future.

Thus, UIDAI was constituted on 28 January 2009 as an attached office under the aegis of the Planning Commission. The UIDAI was, inter alia, given the responsibility to lay down the plan and policies to implement the UID scheme, own and operate the UID database and be responsible for its updation and maintenance on an ongoing basis. It spelt out the functions of UIDAI in detail.[19]

Nandan and I joined UIDAI in July 2009. The work of UIDAI started in August 2009. One of the first things we did was to form two committees. The first was under the chairmanship of N. Vittal, former Central Vigilance Commissioner (CVC), and its mandate was to figure out what demographic data should be collected from residents as part of creation of the unique IDs.[20]

We, at the UIDAI, were acutely aware of the enormity of the 'uniqueness' challenge. It seemed that biometrics were the only tools available for deduplication necessary to ensure that one person does not get more than one ID. Hence, a second committee called Biometric Standards Committee (Biometric Committee), was constituted under the chairmanship of Dr B.K. Gairola, the then DG of NIC, and its job was 'to develop biometric standards that will ensure interoperability of devices, systems and processes used by the agencies that would use the UID system and to review the

---

[19]Notification of constituting UIDAI, https://uidai.gov.in/images/notification_28_jan_2009.pdf. Last accessed on 5 May 2020.
[20]Biometric Standards Committee Notification.

existing standards of biometrics and, if required, modify, extend or enhance these standards to serve the specific requirements of UIDAI relating to deduplication and authentication.' The membership of the committee included the RGI, representatives from banks and academic institutions such as the IITs, the RBI, ICICI and some user groups.[21]

Accuracy of Biometrics Questioned

Even before the constitution of the Biometric Committee, we held workshops, starting as early as September 2009, to investigate the technology and issues with biometrics-based identities. We invited representatives from government departments, academicians, biometric experts and field-level user companies. Generally, the academic community raised doubts about the quality of fingerprint patterns of Indian population and applicability of results obtained with Western data. User community's views were that they had already collected a huge amount of data for their own applications for government programmes such as MGNREGA and PDS. However, they had neither done any testing and analysis of the data nor followed any uniform standards. Therefore, there was no question of interoperability. The international community was of the view that the accuracy of fingerprints could only go up to 90 per cent and not higher.

The first deliverable of the committee was to frame biometric standards based on existing national and international standards, with the consensus of various government stakeholders. The second deliverable of the committee encompassed best practices, expected accuracy, interoperability, conformity and performance in biometrics standards.

As deduplication of the magnitude required by UIDAI had never been implemented anywhere in the world, the Biometric Committee, even as it held discussions with international experts

---

[21] https://uidai.gov.in/images/resource/Biometric_Standards_Committee_Notification. pdf. Last accessed on 5 May 2020.

and technology suppliers, formed a subgroup to analyse the quality of Indian fingerprints. Over 2,50,000 fingerprint images from 25,000 persons were sourced from the districts of Delhi, UP, Bihar and Orissa. Nearly all the images were from rural regions and collected by different agencies using dissimilar capture devices and through different operational processes. Analysis showed that UIDAI could obtain fingerprint quality as good as seen in developed countries, provided proper operational procedures were followed and good-quality devices were used.

The committee found that while deduplication accuracy of 99 per cent had been previously achieved in a database of 50 million, there were uncertainties around retaining this efficacy with a database of a billion individuals. Furthermore, there was evidence to suggest that the quality of fingerprints, and therefore the accuracy of deduplication, drops precipitously if attention is not given to operational processes. As an insurance against possible failure of the fingerprint-based deduplication, and therefore of the project, we advocated the use of iris scans. Our view was that iris as an independent biometric will be able to substantially improve deduplication accuracy.

However, the RGI did not agree with UIDAI's view relating to inclusion of iris scans and favoured using only fingerprints. They expressed their reservations against inclusion of iris in a letter dated 23 December 2009 addressed to Dr Gairola.

The letter pointed to the expenditure already made in collecting 10 fingerprints and a photograph for the Coastal NPR project,[22] which could become infructuous if iris data was also required. It said that inclusion of iris scan for subsequent enrolment in NPR for the whole country would also entail huge additional costs. Citing inadequate data about successfully acquiring iris scans, especially in India, the RGI asked for mandatory biometrics to be restricted to 10 fingerprints and a photograph, leaning on a suggestion made by Professor Phalguni Gupta of IIT Kanpur, a

---

[22]The RGI had collected fingerprints, but no iris scan, of about 28 million people for the MNICs for residents in nine coastal states and four Union territories after the 26/11 Mumbai attack in 2008, under the Coastal NPR project.

member of the committee.

The Biometric Committee, possibly due to a tussle between the RGI, the Department of Information Technology and NIC, was reluctant to say that iris scan should be included for better accuracy. But it was also unable to say that only fingerprint-based deduplication would work reliably at the required scale.

We were prepared for the Biometric Committee to limit itself to prescribing the specifications for all the three biometrics: face, fingerprints and iris. UIDAI would then take the management decision based on the organization's objectives.

Finally, the committee concluded in its report that 95 per cent deduplication accuracy using moderately good fingerprint images, for a database size of one billion, was achievable. Higher accuracy of 99 per cent had been achieved with good-quality fingerprints in a comparably modest database of 50 million. It recommended collecting a photograph and 10 fingerprints as per the International Organization for Standardization (ISO) standards. It also made further recommendations about ensuring process reliability in the collection of biometrics, rewarding enrolling agencies based on the quality of images, ensuring interoperability, vendor independence, etc. It advised that the scans, being national assets, should be preserved in their original quality and resolution. While it did mention that inclusion of iris would improve the accuracy, it left the decision of inclusion of iris to UIDAI.

The Iris Battles

While we failed to get the endorsement of the Biometric Committee, the battle was not over yet. For a robust case for inclusion of iris, we prepared a white paper that detailed the benefits of including iris in the biometric set—the most important being the improvement in accuracy of uniqueness.[23] Other important reasons for iris inclusion were: (i) improved inclusion, (ii) ease of use, (iii) reducing the risk

---

[23]'Ensuring Uniqueness, Collecting Iris Biometrics for the Unique ID Mission.' Available at: https://www.yumpu.com/en/document/read/3700820/ensuring-uniqueness-collecting-iris-biometrics-for-the-uidai. Last accessed on 11 May 2020.

of execution, (iv) reducing technology risk, (v) faster deduplication, (vi) better application development, (vii) security and (viii) future development of ID systems.

We prepared the estimates relating to the incremental cost of iris scans. At about an additional ₹4.20 per enrolment, the total incremental cost for the entire population was going to be roughly ₹500 crore (₹5 billion). We stated that the objective of uniqueness could not be achieved without an iris scan and because UIDAI had the authority to decide the technology to be used in the project, we wanted to take the decision to include the iris scan.

The issue of including iris scan was discussed in several committees and finally put up before the Cabinet Committee on UIDAI on 18 May 2010.[24] This committee approved the adoption of the approach proposed by UIDAI relating to the collection of biometrics, including iris scan. The iris battle was thus finally resolved.

After this decision, we were confident of delivering upon uniqueness, but issues such as speed and scalability of the deduplication process still needed confirmation. Importantly, while uniqueness was mandated by the government, authentication is something we had promised from our end. The Strategy Document presented to the Cabinet Committee on UIDAI in August 2009 (and subsequently updated) promised online authentication of IDs as one of the services of UIDAI.[25]

However, individuals from every section of society—from NGOs to parliamentarians to the academic and scientific community—continued to debate and question the technological feasibility of what we were about to do.

---

[24]'Government gives in principle approval for adoption of uniform and standardised approach for collection of demographic and biometric attributes of residents for UID project', 18 May 2010. Available at: http://pib.nic.in/newsite/erelcontent.aspx?relid=61903. Last accessed on 5 May 2020.

[25]'UIDAI Strategy Overview: Creating a unique identity number for every resident in India'. Available at: https://www.prsindia.org/sites/default/files/bill_files/UIDAI_STRATEGY_OVERVIEW.pdf. Last accessed on 5 May 2020.

Parliamentary Opposition to Identity Bill[26]

The National Identification Authority of India Bill was introduced in Parliament on 3 December 2010 and referred to the Standing Committee, who gave their report on 13 December the following year. While considering the feasibility of technology, among other issues, the committee relied on the testimony of Dr R. Ramakumar, the self-styled expert on biometric technology, who had been consistently writing against the project.[27] He warned the committee that failure to enrol with fingerprints could be as high as 15 per cent in the population dependent on manual labour, thus potentially excluding those who were most in need of the identity.

While Dr Ramakumar's assertions on technology issues such as failure to enrol into Aadhaar being as high as 15 per cent were completely exaggerated, the committee was troubled by the well-intentioned caveats in the Biometric Committee's report and the Planning Commission's admission that nowhere in the world had such a large biometric database of a billion people been created and that the frontiers of technology in biometrics were being tested in the project.

The parliamentary committee naturally balked. It concluded that the Bill in its current form was unacceptable and it was 'unlikely that the proposed objectives of the UID scheme could be achieved.' It, therefore, rejected the Bill, noting contradictions within the Government that related to both implementation and implications. The committee asked that the data collected by UIDAI be transferred to NPR, if the Government so chose, and urged that a fresh legislation be brought before the Parliament.

UIDAI's response to the committee's report was that it had taken steps to ensure that biometrics would yield the required accuracy.

---

[26]Report of the Standing Committee of Finance on the National Identification Authority of India Bill 2010. Available at: https://www.prsindia.org/uploads/media/UID/uid%20report.pdf. Last accessed on 24 July 2020.

[27]Dr Ramakumar has written several newspaper articles, mainly in *The Hindu*, expounding his views on Aadhaar. Some such articles are: 'Aadhaar: Time to disown the idea,' *The Hindu*, 16 December 2011; 'Aadhaar: On a platform of myths,' *The Hindu*, 17 July 2011 and 'What the UID Conceals,' *The Hindu*, 21 October 2010.

Based on the Biometric Standards Committee report, the Proof of Concept (PoC) report, global learnings and expert opinions, UIDAI made the following design choices:

i) Selected three biometric modalities of 10 fingerprints, two irises and face.

ii) Created standard client enrolment software, with quality checks for biometric and demographic data, consistency of capture process and encryption of enrolment data for security and data protection.

iii) An enrolment server that performed demographic deduplication, biometric deduplication and manual adjudication of matches found by the system.

iv) Use of commodity hardware, devices standards and open-source software wherever possible, and defined standards and application programming interfaces (APIs) for interoperability and to avoid vendor lock-in.

By the time the parliamentary committee's report was published, 25,000 active enrolment stations were being operated by 83 active enrolment agencies contracted through 59 active registrars across many states and Union Territories. Each station enrolled 50 residents in a day, on an average. As many as 11 different models of fingerprint and iris devices were deployed in the field. More than 140 million enrolments had taken place using UIDAI's client-side system. Over 100 million Aadhaar numbers had been generated by the server-side system. The throughput had consistently increased since the start of the programme both in the field and at the back end. In November 2011 alone, 20.18 million Aadhaar numbers had been generated and the system was able to process a million Aadhaar numbers per day.

By then, Aadhaar had become the world's largest system in terms of daily processing rate and one of the three largest in terms of its database size. It was the largest multimodal biometric deployment in the world, producing daily measurements of accuracy, throughput and quality that met industry-accepted standards. It had become sufficiently large to accurately and definitively predict the performance necessary to enrol the entire population.

Everyone whom the parliamentary committee had heard, including the government departments concerned, had dithered in giving categorical assurances on the issues the committee had raised. In hindsight, they may have reached a different conclusion if the committee had only given UIDAI an opportunity to respond to its questions and the tentative findings.

When None Is an Expert, Everyone Is!

There were a number of other 'experts' giving views about the technological non-feasibility of biometrics in this project. Dr David Moss, director, Business Consultancy Services Ltd, said that UIDAI's attempt to create unique identities in a population size of 1.2 billion was a pipedream,[28] as it would create an unmanageable 'sea of false positives',[29] going on to assert that there could be up to 15,000 false positives for every Indian resident.[30]

Dr Moss arrived at these conclusions because he had confused false positive identification rate (FPIR) with false match rate (FMR). The former is the probability of a false alarm when a resident is matched with everyone else already enrolled in the system. Whereas FMR is the probability of incorrectly declaring a match in a one-to-one comparison.

To ensure uniqueness, every person needs to be checked for a match with every other person. In other words, for every new enrolment, there is a search in the gallery that involves comparison with all previously enrolled residents. The possibility of a false positive identification in such a search is 0.0025 per cent.[31]

---

[28]David Moss, 'True Lies of Biometric Technology in Aadhaar Enrolment,' Kracktivist, 27 January 2012. Available at: https://kractivist.wordpress.com/2012/01/27/true-lies-of-biometric-technology-in-aadhaar-enrolment/. Last accessed on 5 May 2020.

[29]David Moss, 'India's ID Card Scheme: Drowning in a sea of false positives,' March 2011. Available at: http://www.dematerialisedid.com/BCSL/Drown.html. Last accessed on 5 May 2020.

[30]Moneylife Digital Team, 'How UIDAI Goofed Up Pilot Test Results to Press Forward with UID Scheme,' Moneylife, 18 March 2011. Available at: https://www.moneylife.in/article/how-uidai-goofed-up-pilot-test-results-to-press-forward-with-uid-scheme/14863.html. Last accessed on 5 May 2020.

[31]An alert reader would realize that everything else remaining same, the FPIR must

At 0.0025 per cent FPIR, a million enrolments a day may result in only 25 false positives, which then would need to be adjudicated manually each day. That is hardly unmanageable!

There are hundreds of articles that have been written during the initial years and continue to be written opposing the UID project. A partial list can be found on the blog post 'Say No to Aadhaar'[32] and 'Aadhaar Related Articles'[33].

From a standpoint of technology criticism, the following apprehensions were broadly made to support the assertion that things will not work:

i) It is an untested technology and as the Biometric Committee of UIDAI itself had expressed apprehensions, UIDAI should not go ahead and waste public funds on an untested technology. (The Standing Committee of the Parliament examining the Bill and other experts.)

ii) As there are a large number of people whose biometrics are not of quality due to hard labour, etc., there will be an equally large number of people who will not be able to enrol and hence will be excluded from the system—first at the point of enrolment. And even when you enrol them somehow, they will not be able to authenticate subsequently and will be denied the benefit of various programmes that start using biometric authentication for service delivery. Aadhaar thus becomes a tool of exclusion, especially of the poor and marginalized.[34] (The prominent

---

linearly increase with gallery size. However, the FNIR can be maintained at the target of 0.0025 per cent by a small trade-off with FNIR, or False Negative Identification Rate

[32]'Say No to Aadhaar', Available at: https://saynotoaadhaar.blogspot.com/. Last accessed on 5 May 2020

[33]'Aadhaar Related Articles', Available at: https://aadhaar-articles.blogspot.com/. Last accessed on 5 May 2020.

[34]Pushkar, 'The Shaky Foundation of AADHAR', Green Clean Guide, 3 January 2012. Available at: https://greencleanguide.com/unique-identification-project/. Last accessed on 5 May 2020.

critics in this category are Reetika Khera,[35] Jean Drèze[36,37] and Dr Ramakumar.[38])

iii) The biometric technology at this scale will not be able to ensure uniqueness and will result in a large number of duplicates, thus negating the very purpose of creating unique identities. Some argued that while it may be possible to ensure uniqueness at smaller numbers, it will not be possible when the gallery size becomes large (1.2 billion).[39]

iv) UIDAI's authentication system will not work in a robust manner and false positives may result in fraudulent financial transactions or misappropriation of other benefits.

v) Lack of connectivity will result in authentication systems not working, thus causing exclusions from service delivery.

Testing the Technology

UIDAI did not jump into using the technology. We were also aware that using the technology without adequate confidence through PoC studies would be unwise.

In March–June 2010, UIDAI carried out systematic PoC studies in predominantly rural areas of Andhra Pradesh, Karnataka and Bihar. About 75,000 people were enrolled during the first phase of the PoC study and 60,000 of the same people were re-enrolled during the second phase after a gap of three weeks.

In the study, face photos, iris images and fingerprints of all

---

[35]Reetika Khera, 'UID: From inclusion to exclusion.' Available at: https://india-seminar.com/2015/672/672_reetika_khera.htm. Last accessed on 5 May 2020.

[36]Jean Drèze, 'Dark Clouds over the PDS,' *The Hindu,* 18 October 2016. Available at: https://www.thehindu.com/opinion/lead/Dark-clouds-over-the-PDS/article14631030.ece. Last accessed on 5 May 2020.

[37]Jean Drèze, 'Aadhaar-Based PDS Means Denial of Rations for Many, Jharkhand Study Shows,' The Wire. Available at: https://thewire.in/rights/jharkhand-aadhaar-pds-nfsa. Last accessed on 5 May 2020.

[38]R. Ramakumar, 'Aadhaar Bill: An unconstitutional legislation,' News Click, 14 March 2016. Available at: https://www.newsclick.in/india/aadhaar-bill-unconstitutional-legislation. Last accessed on 5 May 2020.

[39]David Moss, 'True Lies of Biometric Technology in Aadhaar Enrolment,' Moneylife, 27 January 2012. Available at: https://www.moneylife.in/article/true-lies-of-biometric-technology-in-aadhaar-enrolment/23257.html. Last accessed on 5 May 2020.

10 fingers were captured. The fingerprints were captured in two different ways: first, by using a slap device (capturing four fingerprints in one go), and then by using a single-finger device. Rural areas were emphasized in the study because a greater fraction of the rural population is involved in physical labour, whose fingerprints could be worn out. Also, it was necessary to test whether biometric enrolment was feasible in locations that had limited access to electrical power, proper lighting and other support systems.

The objectives of this study were to evaluate technical, operational and behavioural hypotheses related to both the use of biometric devices and the overall enrolment process itself. It was also conducted to establish a baseline for the quality of biometric data that could be collected in rural India.

These studies confirmed that biometric enrolments were possible in Indian conditions and took a little over three minutes, of which less than a minute was spent on iris scan. Many blind people also had their iris images captured. Variation in devices, extremes of weather and the age of those enrolled did not hamper the process, though it did take a little longer to enrol the elderly or those with worn-out hands. Overall, the results were encouraging and indicated that with simple procedural adjustments, the required quality and accuracy could be ensured across the country.[40]

While enrolment PoC established the feasibility of enrolment and deduplication at a small scale, the feasibility and accuracy of deduplication at a large scale could not be established unless we had processed a significant fraction of the total expected enrolments.

In December 2011, when we had reached about 150 million Aadhaar enrolments, we carried out a detailed study to test various parameters including accuracy of biometric deduplication. The gallery size used for this study was 84 million. Following are the highlights of the results:[41]

[40]Planet Biometrics, 'India Releases Biometric Enrolment Report,' 16 December 2010. Available at: https://www.planetbiometrics.com/article-details/i/417/. Last accessed on 5 May 2020.
[41]'Role of Biometric Technology in Aadhaar Enrollment,' 21 January 2012, Page 4. Available at: http://www.dematerialisedid.com/PDFs/role_of_biometric_technology_in_aadhaar_jan21_2012.pdf. Last accessed on 5 May 2020.

As of 31 December 2011, the UIDAI has true and tested statistics computed from real operational large-scale UIDAI system with the resident enrolment database size of 8.4 crore (84 million). It is unnecessary and inaccurate to attempt to infer UIDAI system performance from other systems which are ten to thousand times smaller. Specifically,

• **Failure to Enrol (FTE) Rate: Zero.** As a policy, every unique resident, regardless of their biometrics can be enrolled and issued Aadhaar number.

• **Biometric Failure to Enrol Rate: 0.14 per cent.** This implies that 99.86 per cent of the population can be uniquely identified by the biometric system. The exceptions (0.14 per cent), however, are de-duplicated using demographic data and checked manually for fraud. The legitimate cases among these are issued an Aadhaar number.

• **False Positive Identification Rate (FPIR): 0.057 per cent.** In practical terms, it means that at a run rate of 10 lakh enrolments a day, only about 570 cases need to be manually reviewed daily to ensure that no resident is erroneously denied an Aadhaar number. The UIDAI currently has a manual adjudication team that reviews and resolves these cases. After manual adjudication, there is a negligible number of legitimate residents who are wrongly denied an Aadhaar number.

• **False Negative Identification Rate (FNIR): 0.035 per cent.** This implies that 99.965 per cent of all duplicates submitted to the biometric deduplication system are correctly caught by the system as duplicates. Given that currently approximately 0.5 per cent of enrolments are duplicate submissions, only a few thousand duplicate Aadhaars would possibly be issued when the entire country of 120 crore is enrolled.

The analysis resulting from such a large data set (84 million records) is empirically repeatable and statistically accurate. There is no longer a need to rely on small sample size tests or hearsay from other projects. The UIDAI is now capable of measuring the accuracy, performance and scalability of the actual production system, which is already among the largest in the world. The results lay to rest unfounded claims that the underlying technology is untested, unreliable and based on unproven assumptions.

Based on the analysis, it can be stated with confidence that UIDAI enrolment system has proven to be reliable, accurate and scalable to meet the nation's need of providing unique Aadhaar numbers to the entire population.

One Million Authentications Per Hour

Failure to authenticate[42] should not lead to denial of service. At the same time, it is necessary to assess the percentage of failure of authentication. It was necessary to ensure that people at the field level feel comfortable with the authentication process and exception handling was invoked only in a small percentage of cases.

Therefore, just like for enrolment, UIDAI also carried out detailed PoC studies in both fingerprint and iris authentication. The PoC studies to determine authentication accuracy with fingerprints were done in Karnataka, Delhi, Himachal Pradesh, Maharashtra and Jharkhand in 2011. We studied the impact of factors, such as authentication devices and interoperability, number of fingers, fingerprint quality, network-related parameters and feasibility of buffered authentication on fingerprint-based authentication. These studies helped determine that 98.13 per cent of the population could be successfully authenticated in this manner, with even higher accuracy being achieved with the best devices. It was also demonstrated through benchmarking that the Central Identities Data Repository (CIDR) infrastructure was able to sustain one million authentications per hour.[43]

---

[42]R. Ramakumar, 'Identity Concerns', *Frontline*, Print edition, 2 December 2011. Available at: https://frontline.thehindu.com/cover-story/article30177780.ece. Last accessed on 5 May 2020.

[43]'Role of Biometric Technology in Aadhaar Authentication: Authentication Accuracy – Report', Unique Authentication Authority of India, 27 March 2012. Available at: https://www.yumpu.com/en/document/read/10178448/uidai-role-of-biometric-technology-in-aadhaar-authentication. Last accessed on 5 May 2020.

Similarly, another PoC was done to test the feasibility and accuracy of iris-based authentication in September 2012.[44] This was conducted in Mysore district of Karnataka. About 6,000 residents participated creating about 18,000 authentication transactions. The PoC established that high accuracy can be obtained using both single- and dual-eye cameras, achieving a false reject rate (FRR) of less than 0.5 per cent.

Thus, both the concerns relating to failure to authenticate and accuracy of authentication were set to rest. Of course, now both types of authentications—fingerprint and iris—are being done at the required scale. Starting from attendance in various offices, authentication is being done for lifting PDS ration, getting mobile SIMs, opening bank accounts, accessing bank accounts through Aadhaar-enabled payment system (AEPS), giving digital life certificates and many other applications.

As per the UIDAI's portal,[45] there were around a billion authentications every month in January 2020.

A Showpiece of Technology

We were attentive to the challenge of having to provide unique identities. Having seen duplicate and fake identity papers, such as ration cards or voter ID cards, in earlier years of my service, we were prepared to spend any effort to ensure that we do not fail in this objective of achieving uniqueness. While we had used open-source stack at the back end to architect our CIDR as the data centre, deduplication solutions that were proven and scalable did not exist in the open-source domain.

Therefore, we were constrained to hire an outside agency to conduct biometric deduplication. The United States Visitor and Immigrant Status Indicator Technology, commonly referred to as US-VISIT, is a US Customs and Border Protection (CBP) management

---

[44]Role of Biometric Technology in Aadhaar Authentication: IRIS Authentication Accuracy–PoC Report, Unique Identification Authority of India, September 2012.
[45]Aadhaar Dashboard. Available at: https://uidai.gov.in/aadhaar_dashboard/auth_trend. php. Last accessed on 5 May 2020.

system. It involves the collection and analysis of biometric data (such as fingerprints), which are checked against a database to track individuals deemed by the US to be terrorists, criminals and illegal immigrants. This involves biometric deduplication and matching. UIDAI could engage one of these agencies (described as Biometric Service Provider-BSP) to deduplicate the biometric data, fingerprint and iris. However, there were risks such as vendor lock-in and the issue of determining accuracy, scalability and speed of deduplication that would be delivered. Mashruwala suggested that we could, perhaps, engage multiple BSPs. We discussed the pros and cons and went ahead with this suggestion.[46] Aadhaar was the first project in the world to use multiple BSPs in a plug-and-play style. UIDAI got extremely favourable prices due to competition among the BSPs and finally engaged three.

We incorporated speed, accuracy and hardware requirements in the formula to distribute the work among the three BSPs in the project. The system has been in operation since then, and we also had the occasion to test the plug-and-play architecture when we disengaged one of the BSPs midway. This sufficiently de-risked the accuracy issue in deduplication.

We also started independent checking of deduplication accuracy through Dr A.B. Pandey, our DDG in Mumbai, who was made an independent auditor to do random testing to hunt for duplicates. With the help of a team, he analysed the data and brought out deficiencies and inaccuracies in the deduplication system. We took up this issue with the BSPs and further fine-tuned the system, besides putting in place other methods to continuously hunt for duplicates in our database.

One important aspect is that the Aadhaar database is online. If we cancel an Aadhaar number because we found that it is duplicate, the cancelled number will no more be good for authentication! The holder may hold on to his Aadhaar letter, but it is of no use, as the number is invalid and the Aadhaar back end will refuse to

---

[46]R.S. Sharma, 'UIDAI's Public Policy Innovations,' NIPFP Working Paper, No. 176, 6 September 2011. Available at: https://nipfp.org.in/media/medialibrary/2016/09/WP_2016_176.pdf. Last accessed on 5 May 2020.

authenticate it. Thus, UIDAI can continually mine the data for potential duplicates and cancel the ones it detects. Hence, the accuracy of the database increases over time. Thus analytics, fraud-detection techniques and applications continue to clean the data as we go along.

The current rate of authentications is about 26 million transactions per day, without the imaginary problem of large-scale exclusion due to the failure to authenticate. There have, of course, been scattered cases where incorrect seeding has resulted in service denial to the beneficiary. Rajasthan PDS is an example.[47] However, the fault lay with incorrect seeding of Aadhaar numbers in PDS database rather than Aadhaar authentication.

Diligence, innovative practices, proactive action, open architecture and design for interoperability helped us turn what was once declared a technologically impossible project into a showpiece of technology. Aadhaar has demonstrated the speed, scalability and accuracy in deduplication that were the *sine qua non* for delivering the uniqueness promise. And it has thrown in online authentication as a bonus.

Rather than India looking Westward for technologies, Aadhaar is an example where countries are looking towards India to replicate their identity programmes. Many countries in the world, especially developing countries in Asia, Africa and Latin America have shown interest in learning how India accomplished such a feat.[48] After all, they too have the same problems—leakages due to ghosts and duplicates in their databases.

---

[47] Anumeha Yadav, 'In Rajasthan, There Is "Unrest at the Ration Shop" Because of Error-Ridden Aadhaar,' Sroll.in, 2 April 2016. Available at: https://scroll.in/article/805909/in-rajasthan-there-is-unrest-at-the-ration-shop-because-of-error-ridden-aadhaar. Last accessed on 5 May 2020.

[48] 'The Ripples of India's Big Splash are Now Lapping on Africa's shores,' Identity documents in Africa, *The Economist*, 7 December 2019, page 42.

Chapter 4

INNOVATIONS ON THE GO

Invention, it must be humbly admitted, does not
consist in creating out of void, but out of chaos.

—Mary Shelly, English novelist

In the 1990s, Robert L. Glass published *Software Runaways*, a book in which he draws lessons from 16 spectacular fiascos.[49] It is an interesting and detailed account of large software projects that failed and became multibillion-dollar embarrassments. The author identifies characteristics of projects that fail, what managers tend to do, what works and what doesn't. Much has changed in the way software projects are managed now. They do, however, continue to suffer delays, quality issues, cost overruns and other problems. Denver International Airport's baggage-handling system, presented in *Software Runaways*, had its challenges.

While a missing bag amongst tens of thousands of bags in the airport area can indeed be a problem, a 'missing' human being is a political and technological challenge of an entirely different proportion! Aadhaar, as it concerned human beings, not bags, faced multiple challenges.

There are lessons to learn from failures, as catalogued by Glass. But are there lessons to learn from the success of a project that prevailed against formidable odds? It is a project that applied technology—both at the front end and the back end—besides having to deal with a multitude of issues related to logistics, political and administrative consensus building, implementation challenges in the

---

[49]Robert L. Glass, 'Software Runaways: Lessons learned from massive software project failures.' Prentice-Hall Co. and Yourdon Press. *Communications of the ACM 41.7* (1998): 14–17.

face of connectivity constraints, multiple languages, non-standard addresses and the responsibility for developing an ecosystem of applications.

Fortunately, for those interested in learning from its story, UIDAI was disciplined and active in documenting its systems, processes, ecosystem rules and other efforts that reveal how each battle was fought and every challenge countered, which holds lessons for the future. It has made public a rich repository of documents and other resources, which can be found at UIDAI's portal.[50] Here are some stories that may be helpful.

Primer on Development of IT Solutions

Tom Cargill of Bell Labs famously explained the ninety-ninety rule in software engineering: 'The first 90 per cent of the code accounts for the first 90 per cent of development time. The remaining 10 per cent of the code accounts for the other 90 per cent of the development time.' The software industry has a playbook with pre-scripted strategies to deal with this development model.

When a sizeable opportunity is sensed, the software company parades their top leadership in the client's office. The discussions are about a shared vision and what an IT system could do. If the client gets interested, their sales team takes over the negotiation for the contract. At this stage, there is a minor stumbling block, especially in working with government agencies. The problem is that the government requires contracts with clear deliverables, timelines and the condition that payment will be made after the completion of work. This model doesn't work in procurement of custom-built software because the client is unable to describe the deliverables with required precision. So, a combination of assurances, promises and the procedure for change request (CR)—God forbid, should change become necessary—is instituted. The entire project is divided into several stages and payments are linked roughly to the

---

[50]UIDAI Portal—Documents. Available at: https://uidai.gov.in/resources/uidai-documents.html. Last accessed on 6 May 2020.

effort estimated for completion of each stage. It is all documented in colour laser prints and bound into neat volumes.

Once the contract is signed, everybody is happy that the hard work is over and the project is launched. Now, all that remains is to check off the deliveries against the milestones. Next day, the executives who negotiated the contract get assigned to other projects and the client is handed off to the delivery team. Initially, things go reasonably. The hardware and licenses for the system software arrive right on schedule. The set-up takes a little longer than expected. But of course, both sides agree that the impact on the overall timeline of the project would not be significantly impacted. The first payment also gets a little delayed, but things are yet firmly in control.

Soon enough, milestones for initial deliveries of the actual solution come up. Delivery must be of a system that runs as expected, which means that 100 per cent code must be completed. Even if a missing functionality is only 5 per cent of the project, Cargill's ninety-ninety rule kicks in that the last 10 per cent of the code accounts for 90 per cent of development time. The deadlines stretch. The vendor claims they are bleeding money and starts asking for change requests—powerful words that software companies love and project owners hate. They also seek absurdly long times for what appears to be a relatively simple task. Change requests soon become necessary at every step. Progress happens at a glacial pace. By this stage, the tiger has walked into the cage that was built for it.

It is true that a major problem in most government projects is that we either do not know what we want or do not invest time and effort in describing the details. Often this job is left to the consultants or as it frequently happens in technology-driven projects, the requirements change as development proceeds. Fortunately, at UIDAI, we did not suffer this problem. Our chief technology architect, Dr Pramod Varma, created a detailed software requirement specifications (SRS)[51] document with such precision

---

[51]'Standard Request for Proposals for Providing Application Software Development, Maintenance and Support Services,' Volume II, UID Application Overview and Requirements, 25 January 2010.

that this architecture holds true till date.

I have interacted with a large number of technology people in my career, but never have I come across a person with such brilliance and clarity of thought in software design and solutions as Pramod. Till today, he is associated with UIDAI and I continue to take his help even in the Telecom Regulatory Authority of India (TRAI) for technology projects.[52] Pramod's detailed document clearly defined the scope of work. However, in a project as massive and complex as Aadhaar, you do not have to look for unexpected situations. They surface with regular frequency and not even Pramod could envisage them because they had to do not so much with design, but with execution complexities.

How do you deal with situations like this? Any change, even a minor one becomes a change request (CR), which means you have to pay for it. When CRs raise their ugly head, the solution provider notifies how much each of them would cost and the client or government department informs they cannot conceivably get an approval because it's impossible to certify their cost. While the clock continues to tick, the work gets stalled. The department is unable to change the vendor, and often the timelines for addressing the CRs are such that it causes problems in other dependent modules or areas. We met this fate in Aadhaar too. But for some unconventional play, we may have been strangled by its deathly grip.

UIDAI Breaks the Script

How did UIDAI counter the software industry's playbook to deliver a complex project fast and without any cost overruns? Both UIDAI committees for demographic data and biometrics standards gave their reports quickly (i.e. December 2009), followed by PoC studies for feasibility of data collection, deduplication, accuracy, etc. The details of PoC were drafted while we awaited the development

---

[52]'Public Open Wi-Fi framework: Architecture & Specification (Version 0.5)', TRAI (Telecom Regulatory Authority of India), 12 July 2017. Available at: https://main.trai. gov.in/sites/default/files/Public_Wifi_Architecture_12072017.pdf. Last accessed on 6 May 2020.

of client software for collection of the data. Mindtree was our Application Software Development, Maintenance and Support Services Agency (ASDMA).

Unfortunately, the PoC could not proceed, as the enrolment software was not ready. During reviews, we were told by the ASDMA that the system was being developed, but all our technology team got at each review was a new set of dates. In my mind, the client software was nothing more than an input form with additional facility to capture fingerprints and photograph!

While visiting the RBI headquarters for some meeting in April 2010, during a car journey from the Mumbai airport to the city, I discussed the development of the client software with Nandan. While my view was that there was unnecessary delay, Nandan was more accepting of the need for time. I told him that it had become an activity on the critical path and things were stuck. While Nandan conceded my point that there were some avoidable delays, we hoped things will be resolved soon. The discussion ended there.

To everyone's surprise, Client Software, version 1.0, was presented in the next review meeting and was eventually used for data capture in the PoC. What had happened in the intervening period was that I decided to code the client software myself. I could do it after office hours on four or five weekends.

This had a few positive outcomes for the project. Like an unusual early move in chess neutralizes the knowledge and preparation of the opponent, it broke the vendor's playbook. The people working on technology understood that they couldn't throw jargon at me and the software agency understood that it would be difficult to overstate technology complexities. This probably helped later in the project as I was able to contribute to the technology discussion and my advice was taken seriously!

The habit of coding continued even later when an issue would catch my attention. Nandan would say that writing code was not my job, but I would aver that I was only driven to it by exasperation with the professionals! Later, I got a few resources to work with me on coding the smaller programmes. A team of just two persons, Abhishek Ranjan and Uzair, created many components of the software stack: quality control, enrolment scheduling, packet upload software, etc.

I would wonder whether knowledge of coding was an advantage or a disadvantage in the role as the DG of UIDAI. Was I meddling too much or was there value in combining domain knowledge with that of technology? I do think, however, that getting some components developed by just two persons, outside the work done by the software agency, eventually saved us millions of rupees while speeding up the work.

Innovations in Enrolment

UIDAI's model was to pay the enrolment agencies based on successful generation of Aadhaar for the enrolments done. They paid their operator in a similar way and so the operators tried to enrol as many people in a day as possible. There were complaints that the operators were in a terrible hurry and thus made mistakes in capturing everything—from names, addresses, gender, date of birth to even the photographs—which the enrolled residents noticed when they received the Aadhaar letter. Several little 'innovations' were introduced to overcome the problem of data-entry errors at the time of enrolment.

Dual Screens: Visiting an enrolment site in Karnataka, in Tumkur district, Nandan and I saw that the enrolee waited while the operator made entries of her demographic data. Nandan asked: 'What if the enrolee could also see the data as it was entered?' That led to the introduction of dual screens, one turned towards the person being enrolled. Data-entry errors could now be called out by the enrolee,[53] which helped to substantially reduce errors.

Training and Certification: When enrolment started in mid-2010, the operators sometimes mixed their own biometrics with

---

[53]'Installation and Configuration of Aadhaar Enrolment Client', UIDAI, 27 November 2012. Available at: https://www.nictcsc.com/images/Aadhaar%20Project%20Training%20Module/English%20Training%20Module/module3b_installation_configuration_of_aadhaar_enrolment_client_17122012.pdf. Last accessed on 6 May 2020.

that of the person being enrolled, entered incomplete addresses, didn't capture the photo properly or committed other errors. The metadata helped us understand the behaviour of the operators and the mistakes it caused. So, UIDAI prepared training material and impressed upon the enrolment agencies to organize training for their agents. We also empanelled training agencies whose services the enrolment agency could use,[54] if it did not have the requisite capacity in-house. UIDAI started evaluating the enrolment agencies and their individual operators and published the error rates on a portal. This improved matters, but when an agency terminated the services of an operator whose performance was poor, the latter simply joined another enrolment agency in the same or nearby area. UIDAI then took the decision for certification of operators.[55]

The empanelled certification agencies were to conduct online tests under their proctoring and only then issue the certificates. Sample question papers were prepared and put on the portal and other platforms.[56] Also, the Aadhaar number of the operator was used as her identifier, so a blacklisted operator could not enrol any resident even if he joined some other enrolment agency.

Using PIN Codes to Check Errors: Field names in addresses were an issue, especially in villages where the name of the village or of a habitation within, was the full address. UIDAI also recognized that addresses have different components in different states, such as post office in some versus police station in others. A landmark, such as behind the temple or near the railway station, could also serve as the address sometimes.

A village or even a city could have multiple spellings due to transliteration. Allowing free text entries increased the number of

---

[54]BS Reporters, 'UIDAI Empanels Enrolment & Training Agencies,' *Business Standard*, 21 January 2013, https://www.business-standard.com/article/economy-policy/uidai-empanels-enrolment-training-agencies-110071800032_1.html. Last accessed on 6 May 2020.

[55]'Enrolment Agencies: Personnel Certified and Registered in CIDR Can Go Ahead and Enrol Residents.' Available at: https://uidai.gov.in/ecosystem/enrolment-ecosystem/enrolment-agencies.html. Last accessed on 6 May 2020.

[56]UIDAI Supervisor Exam Practice Demo. Available at: https://www.youtube.com/watch?v=bvb9W0-OwAU. Last accessed on 6 May 2020.

villages from 656 thousand to 783 thousand and made compilation of enrolment data in a geography very difficult. It frustrated the state governments too that needed area-wise percentage of the enrolments, with granularity down to the level of villages.

There are at least three places called Kota, one each in Rajasthan, Karnataka and Uttar Pradesh. But they have different postal codes or PIN codes: 324002, 576231 and 231222. Further, Kota in Rajasthan may refer to the city or to the district, neither of which is congruent with the area identified by 324002. The postal codes, however, have an important property: they uniquely identify a geographical area and all the postal codes together cover the entire habited area of the country served by India Post. In some sense, post codes are to geographical areas as Aadhaar numbers are to residents: both are numbers designed to provide unique identification.

Crowdsourcing for Corrections: When UIDAI decided to use PIN codes as the first entry in the address field, it was discovered that new mohallas (communities) had come up in cities and villages, but India Post's database was not updated. So, UIDAI resorted to limited crowdsourcing, for which we got the PIN code database from India Post and merged into it the city/village/habitation data from Census 2011 records. This database was transliterated in all local languages used in the state or Union Territory concerned. The master list so created was published on the UIDAI portal and a software-driven workflow developed for updating it. When an enrolment agent started work in a locality and found that the name of the habitation did not figure in the master list or that the postal code itself was wrong, he could initiate a request for addition or correction.

The UIDAI regional offices did some basic sanity checks and flagged the requested changes for officers of India Post, typically the postmaster of the post office concerned, to approve. The enrolment agent would see his changes (often the very next day) and could enrol people for that locality. This entire software and workflow was developed in-house.

Smart Enrolment: Error trapping was incorporated in the data-entry forms. For instance, if the enrolee's age was entered as 100, the operator was prompted to reconfirm the information, but an age of 150 was disallowed. Likewise, there were checks or restrictions in other fields, such as for husband's name for a child or a male resident.

The enrolment software contains quality-control safeguards to ensure that the biometrics satisfy the prescribed quality criteria. The resident is allowed a 48-hour window to point out any errors in the acknowledgement slip where her demographic details are printed.

To make it easy and convenient to get enrolled, Uzair Ahmed at UIDAI created an online appointment system, which allowed the resident to locate a centre nearby and book a time slot. It was a simple and effective solution.

Linking Payment with Success: UIDAI's decision to pay after successful generation of Aadhaar numbers itself aligned everyone's incentive into improving quality to avoid failure of enrolments. It also made the enrolment agency and its operators vigilant about residents attempting a second enrolment, which was bound to fail, thus, wasting everybody's time and effort.

Post-Enrolment Innovations

Initially, enrolment packets could be collected from the field on USB sticks and sent to UIDAI Technology Centre after copying the information on hard disks. But this method was neither scalable nor reliable once the pace of enrolment picked up. We tried out a few quick fixes. But they didn't work because operators made many kinds of errors, such as sending duplicate, corrupted or irrelevant packets. Sometimes, they delayed sending the data or failed completely to send a particular enrolment packet. So, a custom client was developed again in-house, at UIDAI.

Dedicated Upload System: The new system ensured that only registered clients could upload enrolment data; duplicates were filtered at the client end itself; the integrity of enrolment packet was maintained and no garbage data could be included in the uploads. Automated client- and server-side queue management took care of unattended uploads even over unreliable connections, retransmission of corrupted packets and correct sequencing of packets, without missing any of them. All of these actions were carried out with end-to-end encryption between the enrolment client and the back end.

The Aadhaar portal complemented the custom Secure File Transfer Protocol (SFTP) client and delivered real-time reports for monitoring resident enrolment and its secure upload into the system. It presented progress reports organized by area (state/district), by registrars or their enrolment agencies, and even by the enrolment client used. We could thus analyse any delays and take corrective action.

This system was developed by our own resource (Abhishek) under the guidance of the technology team, whereas the ASDMA had given a big-ticket CR and a timeline that would have derailed the enrolment process.

Quality Control at the Back End

Some errors could not be automatically trapped at enrolment, such as gender (male recorded as female or vice versa), photograph errors, mixed biometrics (with the operator), age/photo mismatch, etc. If these had gone into the Aadhaar letters, it would have inconvenienced the affected residents and destroyed the credibility of UIDAI.

As enrolments swelled—averaging one million daily—it was impossible to do a quality check of all the enrolments at CIDR. Hence, a workflow-based system was devised on the maker and checker principle. One person will mark the defect in a record and then it will be sent to another one to confirm the error. Once the two agree, the record was sent back to the enrolment agency to correct and submit with appropriate evidence, for which a detailed protocol was devised. The system could scale to take care of enrolments at peak levels.

An interesting source of error was the delay in receipt of Aadhaar letter by the resident. Even with a delivery accuracy of 99.9 per cent, a million letters could be lost or badly delayed in such a large resident population. Assuming that her enrolment hadn't gone through, it could cause the resident to show up for enrolment a second time. The system was designed to reject the second enrolment with a high probability, but it would have involved unnecessary processing of enrolment packets. Therefore, UIDAI added a potential duplicate check at the quality assurance (QA) stage using fuzzy matching of the demographic data. When a match is flagged, the QA operator would check the photographs. If he is unable to make a decision, a biometric match would be carried between potential demographic matches. Only when all these stages are cleared would the enrolment data be submitted for the full biometric deduplication against all previously enrolled residents. A management dashboard showed data-quality scores enrolment agency-wise, pending packets for correction, etc.

Verification of Biometric Exceptions: As biometrics are the basis of ensuring uniqueness, one would think that those who did not have biometrics—usable fingerprints or iris, etc., or even young children who do not have developed biometrics—would be left out. This was unacceptable, because inclusion was one of the overarching goals of Aadhaar. Therefore, UIDAI designed the system so that if the age of the resident/applicant was less than five years, it would not require biometric capture. Similarly, for those adults who fell in the category of biometric exceptions (no hands or iris), the software would not capture biometrics, but had the provision for taking an additional photograph to demonstrate that the person did not have capturable biometrics (photo of hands showing disabled fingers, etc.).

Unfortunately, this provision, which was introduced with a view to include those with biometric exceptions, was occasionally misused. Some enrolment agents colluded and captured the person's photo for the photograph (like a photo scan) and then categorized them as biometric exceptions. It is also possible that such persons existed, but were not physically present for enrolment (maybe they were living outside India).

Once these examples[57,58,59] came to light, we started verification of all enrolments in the biometric-exception category. It was possible to do so because their numbers were small. Where fraud was detected, UIDAI cancelled the Aadhaar numbers, thus making it impossible to authenticate with those numbers, even if the Aadhaar letter was in somebody's possession.

Innovation Post-Aadhaar Generation

Managing the enrolment ecosystem and processing systems at CIDR was a complex task, but post-Aadhaar generation activities were no less challenging. As the rate of generation of Aadhaar numbers grew, UIDAI had to print up to two million letters every day and ensure their timely delivery. There were delays due to errors in the address, although earlier measures to standardize the address format, training of operators and data quality check through QA helped contain the problem to an extent. It was, however, a mammoth logistical exercise.

Enrolment, Aadhaar generation, printing of letters and their delivery to the resident are sequential activities. Therefore, bottlenecks at any stage created issues that carried over to the last one. And it made the residents restless, and sometimes angry, if their letters didn't reach them on time. UIDAI partnered with India Post for printing and delivery of letters. Once Aadhaar is generated, the XML files[60] are sent to the print partners—hired

---

[57]Mahesh Buddi, 'Aadhaar Card Scam Unearthed in Hyderabad', *The Times of India*, 27 April 2012. Available at: https://timesofindia.indiatimes.com/india/Aadhaar-card-scam-unearthed-in-Hyderabad/articleshow/12888048.cms. Last accessed on 6 May 2020.

[58]Mahesh Buddi, 'Seven Booked in Aadhaar Fraud', *The Times of India*, 17 June 2012. Available at: https://timesofindia.indiatimes.com/city/hyderabad/Seven-booked-in-Aadhaar-fraud/articleshow/14190401.cms. Last accessed on 6 May 2020.

[59]'Team Held in Hyderabad for Issuing Fake Aadhaar Cards'. Available at: https://www.youtube.com/watch?v=Q8U8fnmPBTs. Last accessed on 6 May 2020.

[60]Extensible Markup Language (XML) is a markup language that defines a set of rules for encoding documents in a format that is both human-readable and machine-readable.

by India Post—to print the letters, arrange them by PIN code and hand them over to India Post for delivery.

This simple process ran into three problems: the first was printing of the same letters multiple times by the print partner, resulting in delivery of multiple Aadhaar letters to the same resident. It led to disputes about payment for services rendered. While UIDAI counted the Aadhaar numbers generated, India Post would expect to be paid for the number of letters printed and delivered.

The second problem was non-delivery of letters to a large number of residents. The printer would print letters by PIN code and thus flood the postman working in that area, who was ordinarily geared up to deliver far fewer letters. Soon, stories of Aadhaar letters being dumped in garbage bins or ponds started appearing in the press. This was disturbing, but an attempt to resend would have only aggravated the problem.

The third problem was in reconciliation of delivered letters. Speed Post was chosen because it has a system of recording the delivery of the letter. When Aadhaar letters were handed over to India Post, a file was created that had details as a list. This file was expected to be updated upon delivery. However, this did not happen and UIDAI had no reliable data about delivery failures.

To overcome these challenges, we modified the Print Batch to generate letters of same or similar languages in different PIN codes. Their bundling was organized such that the number of letters in each post office did not exceed 3,000 in a window of 15 days. The smaller-sized lots helped ease the burden of delivery for the postmen.

A solution was also devised to deduplicate every letter that was going out against the records of letters already sent and the printers were given unique files with 600 records each. A robust reconciliation mechanism was instituted for print files with vendors (the first problem) and delivery acknowledgements with India Post (the third problem) to ensure that each letter was delivered and accounted for.

e-Aadhaar: As Aadhaar is an online digital identity, a physical print does not have much value, except that the letter bears the number and other details on it. However, non-delivery of this letter created a demand for another copy. More importantly, people were not even aware of the status of their Aadhaar letter. Therefore, we introduced the facility for checking the status of Aadhaar enrolment online and locally printing a copy of the letter after providing a few details such as the enrolment ID or Aadhaar number and PIN code. This printout was as valid as the Aadhaar letter issued by UIDAI. e-Aadhaar became an instantaneous hit and solved the problem of non-delivery of letters. Again, this was not developed by ASDSMA, but by one of our in-house programmers.

Checking Duplicate Aadhaar Issuance: Biometric deduplication and matching is not perfect. Though the error rate is less than 0.01 per cent, the error count could be large if the whole population made multiple attempts at enrolment. So, we took some additional measures.

UIDAI started independent checking of deduplication accuracy through random testing to filter out duplicates. UIDAI took up the issue of deficiencies and inaccuracies of the deduplication system with the Biometrics Solutions Provider (BSP) and further fine-tuned the system.

As the Aadhaar database is online, it is possible to continuously churn this database and check for duplicates, which can be cancelled after confirmation—using photos and biometrics—with intimation of cancellation to the person concerned. Continuously mining the data increases the accuracy of the database over time. Also, in subsequent authentications, inaccuracies get discovered through analytics or fraud detection techniques, which also helps to clean the data.

Aadhaar was a complex project with the interplay of technology, public policy, inter-agency coordination and logistics at unprecedented scales. During implementation, many challenges arose, which could not have been anticipated. Creating and rolling out solutions to these problems required ingenuity, understanding of technology by the project team, speed and scalability. Lastly, deployment of solutions in the project was like changing the tyres of a running vehicle.

While handling these issues might have been possible within the traditional model of software development lifecycle (SDLC), the slow and halting speed of such a process would have caused failure of the project itself. We could overcome the challenges only because we had an ingenious team of software designers and programmers, who would address the issues as and when they arose.

Section II

COALITION AGAINST AADHAAR

Chapter 5

CONCERN FOR RIGHT TO PRIVACY

I don't reject caution, but you have to be careful about caution because there's a stage when it turns into paralysis.

—Yair Lapid, Israeli journalist-turned-politician

Liberal intellectuals and activists are society's conscience keepers: they do not hesitate to deliver a sharp rebuke where it is warranted. They help us stay on the approved path. They keep us safe and together. So, in a sense, much like the sheepdog, they are a part of the system. They feel responsible towards their less-fortunate brethren in society because of the privilege they themselves enjoy. They are not subversive and don't seek to upend established systems that have served us well.

Aadhaar, although a government project, is subversive. It challenged conventional wisdom on how ID systems were done. It imagined changes to the established ways in which many of our systems worked. And it was powerful enough to have us all worried. Naturally, therefore, Aadhaar received intense scrutiny, especially regarding its impact on the individual's right to privacy.[61,62]

---

[61]Mohul Ghosh, 'Aadhaar Card Bill Will Bring Exceptional Benefits; But, Privacy Remains a Concern,' Trak.in, 17 March 2016. Available at: https://trak.in/tags/business/2016/03/17/aadhaar-ccard-benefits-privacy-concerns/. Last accessed on 6 May 2020.

[62]Amber Sinha and Pranesh Prakash, 'Privacy Concerns Overshadow Monetary Benefits of Aadhaar Scheme,' Hindustan Times, 12 March 2016. Available at: https://www.hindustantimes.com/india/privacy-concerns-overshadow-monetary-benefits-of-aadhaar-scheme/story-E3o0HRwc6XOdlgjqgmmyAM.html. Last accessed on 6 May 2020.

Online publications such as Scroll.in[63] and The Wire[64] regularly featured stories criticizing Aadhaar, especially on the grounds of privacy issues. But maximum criticism against Aadhaar, however, has come from civil society groups.[65,66] Several public interest litigations (PILs) were filed before the Supreme Court challenging Aadhaar on various grounds, including breach of privacy.[67] However, Aadhaar's likely impact on privacy wasn't debated in public discourse and the courts alone, but rather started within UIDAI itself. Nandan considered the issue as central to the development of the unique ID system.

Many of us were in agreement with this view, though the career bureaucrat in me wondered if we were not trying to slap on an elitist and Western concept of privacy onto the new ID system. Would it damage the utility of the instrument even before it was given birth? I was prepared to be damned rather than let theoretical niceties diminish a system sorely needed for the people, especially the poor and marginalized of our country.

I grew up in rural UP and spent my formative years as a bureaucrat in Bihar and Jharkhand. The India I had seen had a weak notion of privacy. Even strangers, in casual conversation, freely ask questions unthinkable in the West. How was your wedding arranged? How many children do you have? When do you plan to get them married? To which caste or sub-caste, religion or political group do you belong? Do you approve of others who eat non-

---

[63]Saikat Datta, 'The End of Privacy: Aadhaar is being converted into the world's biggest surveillance engine', Scroll.in, 24 March 2017. Available at: https://scroll.in/article/832592/the-end-of-privacy-aadhaar-is-being-converted-into-the-worlds-biggest-surveillance-engine. Last accessed on 6 May 2020.

[64]Jean Drèze, 'Hello Aadhaar, Goodbye Privacy', The Wire, 24 March 2017. Available at: https://thewire.in/government/hello-aadhaar-goodbye-privacy. Last accessed on 6 May 2020.

[65]'UID Research', The Centre For Internet & Society. Available at: https://cis-india.org/internet-governance/blog/uid-research. Last accessed on 6 May 2020.

[66]Usha Ramanathan, 'A Shaky Aadhaar', The Indian Express, 30 March 2017. Available at: http://indianexpress.com/article/opinion/columns/aadhar-card-uid-supreme-court-a-shaky-aadhaar-4591671/. Last accessed on 6 May 2020.

[67]'The Aadhaar Case', The Centre For Internet & Society, filed by Justice Puttuswamy. Available at: http://cis-india.org/internet-governance/blog/the-aadhaar-case. Last accessed on 6 May 2020.

vegetarian food? How well off are you financially? All such questions are par for the course. There is neither any offence taken, nor such conversations considered an intrusion into a person's privacy. A blunt refusal to answer, on the other hand, may be taken as stand-offish behaviour.

Such 'meddling' in private affairs is not restricted to the social context alone. In rural India, to a varying degree—depending upon the region, the state itself is fairly intrusive. Indeed, anywhere in the country, a policeman may stop a citizen, and on the basis of suspicion alone, ask him or her to provide a name and address. A refusal could lead to arrest under the Code of Criminal Procedure, 1973.

So, even as I prepared to join UIDAI, I questioned the usefulness of privacy as a key idea around which to design the new ID system, both in my mind and also in discussions with others. Debates revolved around the sanctity of the right to privacy of the individual, which would later play out in the media and the courts. It is a credit to Nandan's leadership and the diligence of the founding team members who analysed the issues and took decisions that would later survive the severest challenges thrown at it.

The sine qua non to Aadhaar

Despite the absence of formal legislation on privacy in India, UIDAI informed itself of privacy concerns and the work done on this subject around the world. People generally confuse privacy of the individual's data with data security. Terms such as data security, data protection and data privacy, although related, are not the same and their usage has often created confusion.

While data security refers to protection of data from unauthorized access and data corruption throughout its lifecycle, data privacy is broadly defined as the appropriate use of data, in accordance with the wish of the individual to whom it pertains and with her consent. Therefore, data privacy is the right of an individual to have control over how personal information is collected and used.[68] Hence, data

---

[68]'What Does Privacy Mean?' International Association of Privacy Professionals (IAPP).

privacy is focused on the use and governance of personal data—like the policy to ensure consumers' personal information is collected, shared and used in appropriate ways.

Data security focuses more on protecting data from malicious attacks and the exploitation of stolen data for profit. Obviously, privacy too can be affected by a security breach. In that sense, these two concepts are closely related, though not identical. Data security practices typically include measures such as data encryption, tokenization and key management practices that protect data across all applications and platforms. While security is necessary for protecting data, it's not sufficient for addressing privacy.

The UIDAI leadership's commitment to privacy is demonstrated by the fact that its chairman, as early as in May 2010, had written to the PM, affirming that while UIDAI will do everything within its control to ensure data security and protection, there was a need for a general and omnibus data protection and privacy law in the country. Its own proactive steps towards that aim later came to be incorporated in the Aadhaar Act itself.

In its Strategy Overview document, UIDAI declared that it envisioned:

> [A] balance between 'privacy and purpose' when it comes to the information it collects on residents. The agencies may store the information of residents they enrol if they are authorized to do so, but they will not have access to the information in the UID database. The UIDAI will answer requests to authenticate identity only through a 'Yes' or 'No' response.[69]

The Strategy Paper, in the section titled 'Protecting Privacy and Confidentiality,' stated:

Available at: https://iapp.org/about/what-is-privacy/. Last accessed on 6 May 2020.
[69]'UIDAI Strategy Overview: Creating a unique identity number for every resident in India,' page 4. Available at: https://www.prsindia.org/sites/default/files/bill_files/UIDAI_STRATEGY_OVERVIEW.pdf. Last accessed on 6 May 2020.

The information that the UIDAI is seeking is already available with several agencies (public and private) in the country. The only additional information being sought by the UIDAI are fingerprints and iris scans. However, the UIDAI recognizes that the right of privacy must be protected, and that people are sensitive to the idea of giving out their personal information, particularly when information being stored in a central database, will be used for authentication. UIDAI will protect the right to privacy of the person seeking the unique identity number. The information on the database will be used only to authenticate identity. Necessary provisions would be in place to address the issues of privacy and confidentiality.[70]

UIDAI recognized the potential risks in the area of privacy and put in place the mechanism to deal with these risks. To protect the security and privacy of resident data collected during the enrolment process, access to authentication services are given only to authorized ecosystem partners of UIDAI. Under no scenario, biometric data of residents is shared. Further, UIDAI deployed robust security infrastructure and put in place appropriate encryption mechanisms to prevent any unauthorized dissemination of demographic or biometric data of residents stored in the CIDR.

Additionally, as the Aadhaar system continually interacts with other databases for seeding and authentication, it is backed by the Aadhaar Act, which prohibits certain activities that could otherwise dilute privacy and data protection. Therefore, the design of the system itself, its operation and supporting legal provisions promote privacy and data security as outlined below.

Minimal Data Collection: The principle is that you must start with a null set, i.e., the assumption that you need to collect and store no information. To this set, you may add such data elements that are absolutely necessary for the purpose you want to achieve and justify every addition to the list with the rational nexus that exists between data sought to be collected and the objective of the project.

---

[70]ibid.: 32.

The purpose of Aadhaar is to establish the identity of an individual that is unique. Therefore, the following question was put to the Demographic Data Committee: what are the essential attributes of the identity—the demographic information—and why must they be collected?

The committee, in its report[71] recommended that only four demographic attributes of the identity were essential for the objectives of the programme, viz. name, date of birth, gender and communication address. The committee turned down suggestions to include attributes such as place of birth, permanent address, income, etc., as it considered these attributes not essential for the purpose of establishing identity.[72] UIDAI in its Strategy Document specified that it will collect only basic information from the resident.

Although the four types of demographic information—name, date of birth, gender and address—would serve to establish an individual's identity, it was not enough to prove uniqueness. These were necessary but not sufficient. Therefore, the biometric information of an individual was added as an essential attribute for ensuring uniqueness. On the issue of biometric data, Aadhaar collects photograph and images of both iris and all 10 fingerprints. Photograph is certainly an essential data for establishing identity, but the technology adopted by Aadhaar uses iris and fingerprints as core biometric data that is used to guard against a duplicate before an identity record is added to the database. These technology choices were not made trivially.

UIDAI also optionally collects mobile number and email id, the purpose being to communicate with the resident for any activity such as authentications, tracking data updates, etc. Mobile number becomes not only a communication tool but also an alternative way of authentication, once it is linked to Aadhaar.

The Aadhaar Act[73] specifies that UIDAI will be able to collect

[71]Demographic Data Standards and Verification procedure (DDSVP) Committee Report, UIDAI, December 2009. Available at: https://uidai.gov.in/images/UID_DDSVP_Committee_Report_v1.0.pdf. Last accessed on 6 May 2020.
[72]ibid.: 7.
[73]Section 2(k) of the Aadhaar Act.

some information relevant for the purpose of issuing an Aadhaar number. However, it prohibits collection of information such as race, religion, caste, tribe, ethnicity, language, records of entitlement, income or medical history. This ensures that the data collected is sufficient for establishing the identity of the person and cannot be used for any other purposes, even in the aggregate. Thus, the principle of purpose specification is satisfied.

Restrictions on Data Usage: UIDAI declared in its Strategy Document that the data collected will only be used for issuing the Aadhaar numbers, and later, for providing authentication service to the residents. Specific consent was also taken from the residents if they were willing to share demographic data with the bank for opening a bank account.

The Act[74] also has provisions relating to limitation of usage of information available with UIDAI. As a further safeguard, the biometric information, as per section 29(1)(b) of the Act, shall not be 'used for any purpose other than generation of Aadhaar numbers and authentication under this Act.'

Keeping Residents Informed: While the Strategy Document does not specify that residents will be informed about the use of their data at the time of enrolment, consent relating to usage of data is taken, especially for demographic data to be shared with the bank for opening an account. Section 3(2) of the Aadhaar Act also mandates that residents shall be informed about the manner of use of the information, details of sharing of information and have the right to access their own information.

The authentication process, as defined in the Strategy Document as also by law, implies that the owner of the data—the resident— must participate in the process of authentication. At the same time, Section 8 of the Act explicitly provides that the entity requesting authentication shall be responsible for informing the resident and also to get consent about the authentication. Hence, the resident is always kept informed of the purpose, nature of information to

---

[74]Section 29(3) of Aadhaar Act.

be shared during authentication and the use of the information received during this process.

Random Number with No Intelligence: Another privacy-driven decision was to use a random number for Aadhaar. Section 4(2) of the Aadhaar Act also mandates the same. The number does not disclose anything about the Aadhaar holder. It does not disclose the gender, age or any other attribute relating to one's identity. By examining the number itself, including its validation by way of the check digit, one cannot determine if a given number that looks like Aadhaar is actually one or not!

Stringent Data-Sharing Policies: UIDAI adopted a strict data-sharing policy: no data download is permitted and search is not allowed on any attribute. To quote from the Strategy Document:

> **The UIDAI will not share resident data:** The UIDAI envisions a balance between 'privacy and purpose' when it comes to the information it collects on residents. The agencies may store the information of residents they enrol if they are authorized to do so, but they will not have access to the information in the UID database.[75]

There are only a few ways in which one can interact with the Aadhaar database. One of these is authentication, the process by which the Aadhaar number, along with other attributes (demographic/biometrics/one-time password [OTP]) is submitted to UIDAI's CIDR for verification. CIDR verifies whether the data submitted matches the data available in CIDR and responds with a 'yes' or 'no'. No personal identity information is returned as part of the response. Furthermore, it is only the third-party authentication user agencies (AUAs) who can submit such requests.

Data can also be shared based on an explicit and voluntary authorization from the owner of the data, i.e., the Aadhaar holder. This is for the process of eKYC—a service that enables a resident to

---

[75]'UIDAI Strategy Overview: Creating a unique identity number for every resident in India,' page 4. Available at: https://www.prsindia.org/sites/default/files/bill_files/UIDAI_STRATEGY_OVERVIEW.pdf . Last accessed on 6 May 2020.

share their demographic information and photograph with a UIDAI partner organization in an online, secure and auditable manner. The consent by the resident can be given via a biometric authentication or through OTP authentication.

In eKYC, the Aadhaar holder does a biometric or OTP authentication with the request that her demographic data and photograph be shared (eKYC) with the requesting agency. After the authentication attempt is successful, the authority releases a digitally signed document called eKYC, which can be used by the Aadhaar holder as an ID proof to the service-delivery agency such as a bank or a mobile company. The biometrics are not shared in any of these processes, except in certain situations such as national security or under the orders of a competent court.

Chapter VI of the Aadhaar Act deals with the subject of Protection of Information and Section 28 casts the responsibility of data protection upon the Authority. Section 28(1) & (2) state that 'the Authority shall ensure the security of identity information and authentication records of individuals' and 'subject to the provisions of this Act, the Authority shall ensure confidentiality of identity information and authentication records of individuals.'

Biometric information is also given special treatment in this Act. Section 30 declares biometrics to be 'sensitive personal information' within the meaning of the Information Technology Act, 2000. As per Section 43A of this Act, 'sensitive personal data or information' means such personal information as may be prescribed by the central government in consultation with such professional bodies or associations as it may deem fit.

There are stringent provisions governing bodies entrusted with handling sensitive personal data. If such a body 'is negligent in implementing and maintaining reasonable security practices and procedures and thereby causes wrongful loss or wrongful gain to any person, such body corporate shall be liable to pay damages by way of compensation, not exceeding five crore rupees, to the person so affected.'[76] This section effectively makes UIDAI liable for

---

[76]Section 43A of the Information Technology Act, 2000.

damages and compensation in case it does not perform the duties of data protection cast upon it and the loss of information occurs.

While adequate safeguards have been provided in the Aadhaar Act relating to safety, security and protection of data, the Act does make an exception in two specific cases: national security and judicial orders. However, even these exceptions have requisite safeguards and these provisions have been further strengthened by the order of the Supreme Court and the Ordinance issued by the government thereafter.[77]

Federated Data Model: UIDAI does not keep any data except the logs of authentication executed by the Aadhaar holder. It does not know the purpose of authentication. It only knows the date, time and the agency through which the authentication was done. Thus, the transaction details remain with the agency concerned and not with UIDAI. This federated data model keeps the data with the relevant data controller, who has the responsibility for data confidentiality and security. This principle is articulated in Section 32(3) of the Aadhaar Act that stipulates that 'the Authority shall not, either by itself or through any entity under its control, collect, keep or maintain any information about the purpose of authentication.' This provision ensures that Aadhaar does not facilitate aggregation of information about an individual or a population.

End-to-End Encryption: Aadhaar enrolment is done by the Registrars of UIDAI through enrolment agencies. Enrolment agencies are typically private agencies, who could pose a risk of breach of data. However, this possibility has been eliminated by first ensuring that enrolment is done through a standardized piece of software and that data is encrypted at the time of enrolment itself though an encryption key of 2048 bits. Thereafter, the data is kept encrypted all the time: during transit and at CIDR. It is momentarily decrypted at the time of processing only, which has ensured that there has not been even a single instance of data breach from the UID system. Even when the enrolment machine

---

[77]The Aadhaar and Other Laws (Amendment) Ordinance, 2019.

with data on it is stolen or misplaced, the data cannot be misused, as it is encrypted. This is elucidated in the Aadhaar Technology Architecture document that was established in the early days of the project and exists for public viewing on UIDAI's website.[78]

Furthermore, Aadhaar numbers that are seeded in the systems of third-party operators or AUAs are themselves protected—or ought to be—by standard methods and protocols used in the digital world. The AUAs are typically large entities that have the requisite infrastructure to defend their systems, which they need to do not only to protect the Aadhaar numbers they have collected, but their own digital resources too. The data protection laws, the Supreme Court's recent judgement[79] and the new data protection draft bill all seek to make this responsibility clear and unambiguous.

User in Charge: UIDAI has also built a facility wherein you can 'lock' your Aadhaar[80] and disable it from any type of authentication for a period of your choice. This is an additional measure for securing privacy and confidentiality. Once locked in this manner, biometric authentication is disabled for that Aadhaar number, thus preventing any use.[81] The Aadhaar number holder too, has access to identity information only and not biometric information as per Section 28 of the Act. Similarly, Section 32 of the Aadhaar Act mandates UIDAI to keep authentication logs of the residents, to which the resident may request access.

After each authentication attempt, whether it fails or is successful, UIDAI sends a message on the resident's mobile as well as email—if available—much like the messages sent by banks confirming customers' card transactions. This is an additional

---

[78]UIDAI Aadhaar Technology Architecture, March 2014, Page 20. Available at: https://archive.org/details/Aadhaar-Technology-Architecture/page/n3/mode/2up. Last accessed on 6 May 2020.

[79]Supreme Court Writ Petition (Civil) No. 494 of 2012 in the case of Justice K.S. Puttaswamy (Retd) vs the Union Of India on 26 September 2018.

[80]https://uidai.gov.in/media-resources/resources/videos/12119-lock-unlock-aadhaar. html. Last accessed on 6 May 2020.

[81]'Lock/Unlock Biometrics', UIDAI. Available at: https://uidai.gov.in/contact-support/have-any-question/925-faqs/aadhaar-online-services/biometric-lock-unlock.html. Last accessed on 6 May 2020.

protection for the user to be alerted if any attempts are made to falsely authenticate their identity. It provides comfort, much as a motion sensor does in one's porch or backyard.

Privacy by Design (PbD)

These practices satisfy the end-to-end security and life-cycle protection of resident data, which is one of the important principles of PbD—one of the many recognized approaches to data protection and privacy.[82] The 32nd International Conference of Data Protection and Privacy (Jerusalem, 2010) recognized PbD as the preferred approach to privacy protection. It is an approach to systems engineering which takes privacy into account throughout the engineering process.

The concept of PbD is closely related to that of 'privacy-enhancing technologies' or PETs, a term first used in 1995[83] to describe protection of informational privacy by eliminating or minimizing personal data, thereby preventing unnecessary or unwanted processing of personal data, without the loss of the functionality of the information system.[84]

One of the original proponents of PbD was Ann Cavoukian, information & privacy commissioner, Ontario, Canada. She laid down seven foundational principles for achieving the desired goal.[85] It views that the future of privacy cannot be assured solely by compliance with regulatory frameworks; rather, privacy assurance must ideally become an organization's default mode of operation.

The United Kingdom's (UK) Information Commissioner's Office (ICO), an independent authority set up to uphold information rights

---

[82]Different approaches have been adopted by various countries for dealing with the issue of data protection and privacy.

[83]A. Cavoukian and G. Hustinx, 'Privacy-Enhancing Technologies: The path to anonymity,' Privacy Commissioner of Ontario, Canada (1995).

[84]G.W. Van Blarkom, John J. Borking and J.G. Eddy Olk, 'Handbook of privacy and privacy-enhancing technologies,' Privacy Incorporated Software Agent (PISA) Consortium, The Hague 198 (2003).

[85]Ann Cavoukian, 'Privacy by Design: The 7 foundational principles, implementation and mapping of fair information practices,' Information & Privacy Commissioner, Ontario, Canada. Available at: https://iab.org/wp-content/IAB-uploads/2011/03/fred_carter.pdf. Last accessed on 6 May 2020.

in the public interest, also adopted PbD as 'an approach to projects that promote privacy and data protection compliance from the start'. The ICO encourages organizations to ensure that privacy and data protection is a key consideration in the early stages of any project and then throughout its lifecycle. It is seen as an essential tool in minimizing privacy risks and building trust.

Does Aadhaar Comply with PbD Norms?

UIDAI's leadership articulated its vision to make privacy central to the programme's design and took actions consistent with this stand right from the early days. Aadhaar thus complies with the first principle of PbD that promotes proactive adoption of strong privacy practices, early and consistently, with a commitment at the highest levels towards privacy protection.

The second foundational principle declares 'Privacy as the Default' and lays down basic principles relating to the collection and use of personal information. As one of the objectives of Aadhaar was to help eliminate duplicates and ghosts in the target database, it permits seeding of linked databases with Aadhaar numbers—but it allows nothing more. To minimize any risk from this seeding, the Aadhaar Act prohibits publishing of the linked records with the UID numbers.

The third principle, 'Embedding Privacy in the Design' requires that privacy must be embedded into technologies, operations and information architectures in a holistic, integrative and creative way and is not bolted on as an add-on. Random number for Aadhaar, data-sharing policies and Federated Data Model (FDM) are some of the operational practices that contribute to this principle.

The fourth principle is essentially about ensuring that privacy protection does not mean sacrificing functionality in any way. It asserts that PbD avoids the pretence of false dichotomies, such as privacy vs security, demonstrating that it is possible to have both. It is a positive sum and not a zero-sum situation.

The uniqueness feature of Aadhaar ensures removal of duplicates and ghosts from databases into which it is seeded. Authentication services of UIDAI allow identification of resident without revealing any information in the process. It is the resident himself or herself who may reveal information to the agency for a service it provides. So, UIDAI indeed fulfils the promise of positive-sum and not zero-sum.

Even in the eKYC service, similar to authentication, the user authorizes UIDAI to release identity data to a given entity such as a bank or mobile company. This process too does not dilute any of the privacy principles and the data sharing happens with the user's consent and authority only.

The fifth principle of 'End-to-End Security—Full Lifecycle Protection' requires that information be kept encrypted at every stage. Not only has this been UIDAI's practice since the start of the project, but it was even incorporated into the law[86]. The enrolment data is encrypted upon collection and decrypted only momentarily for processing purposes, such as when BSPs are allowed 'read' permission only for biometric deduplication. The transit data— authentication requests, authentication response or eKYC—always travels encrypted. The law casts responsibility on UIDAI to employ the best technologies and practices to ensure this security.

The sixth principle of visibility and transparency in the operation of the system is honoured by informing residents at the time of enrolment about the use of the collected information and their right to access it. Consent is required at the time of authentication and the user has the option to lock or unlock the biometrics at any time.

The last principle of PbD relates to respect for user privacy. This is clearly in evidence in UIDAI's design of its systems and processes. Its project architects and operators are required to keep the interests of the individual uppermost with user-centricity being the operating principle.

An Evolving Right to Privacy

As the debate revolves around the sanctity of the individual's right to privacy, it is imperative to look at legal status and nature of this

---

[86]Section 28 of the Aadhaar Act.

right in the Indian context. Is it an ordinary civil right or does it enjoy greater protection by the Constitution as a fundamental right? Although, the right to privacy does not find an explicit mention in the Constitution of India, this law has evolved in India through various judicial pronouncements over the years, with a definitive pronouncement being made by the Supreme Court only in 2017.[87]

In 2008, the government had tried to address the legislative deficit by inserting Section 43A into the Information Technology Act, 2000,[88] which relates to compensation for failure to protect data. Under this provision, if a body corporate that possesses, deals with or handles any sensitive personal data or information, is found to be negligent in implementing or maintaining 'reasonable security practices and procedures' in dealing with such data, would be liable to pay compensatory damages to anyone who has suffered wrongful loss or gain as a result.

Section 43A, however, only addresses one part of a much larger issue. There is no actual legal framework to determine the requirements for data quality or data transparency that properly addresses data-protection issues as per accepted norms and standards. Neither is there a data-protection authority to oversee its implementation and test the adequacy and proportionality of actions mandated or taken by various agencies.

In 2011, the government notified eight rules[89] under Section 43A that enacted concepts and principles that privacy lawyers normally expect to see in a data-protection law—definitions of 'personal information' and 'sensitive personal information,' the obligation for prior consent and the requirement to limit the purpose for which data was collected besides restrictions on its transfer and for how long it is needed to be retained. But all these legislative measures

---

[87]Justice K.S. Puttaswamy (Retd) and Anr. vs the Union of India and Ors. (W.P. (Civil) No. 494 of 2012).

[88]Section 43A of the Information Technology Act, 2000. Available at: https://indiankanoon.org/doc/76191164/. Last accessed on 6 May 2020.

[89]Ministry of Communications & Information Technology, Dept of IT, Notification of 11 April 2011: The Information Technology (Reasonable security practices and procedures and sensitive personal data or information) Rules, 2011. Available at: https://www.wipo.int/edocs/lexdocs/laws/en/in/in098en.pdf. Last accessed on 6 May 2020.

were formulated after the birth of UIDAI.

In 2011, a draft bill on right to privacy[90] was prepared by the government, but it could not be turned into a statute. Thereafter, in July 2017, the government constituted a Committee of Experts[91] under the chairmanship of the former Supreme Court judge, Justice B.N. Srikrishna, 'to study various issues relating to data protection in India, make specific suggestions on principles underlying a data protection bill and draft such a bill.'

In July 2018, the committee submitted its report on the data-protection law that contains recommendations for data privacy and security. The report states that personal data shall be processed only for purposes that are clear, specific and lawful. Individuals will have the right to withdraw their consent. Data-protection officers must be appointed to address individual grievances on privacy matters. The Centre shall appoint a Data Protection Authority of India (DPA) as an independent regulatory body as also an appellate tribunal responsible for the enforcement and effective implementation of the law. The law will not have retrospective application and it will come into force in a structured and phased manner.

The draft bill cleared by the Cabinet on 4 December 2019 was introduced in Parliament (Lok Sabha) on 11 December. The Bill has been referred to a Joint Select Committee composed of parliamentarians from both the Lower and Upper houses.

Test of Aadhaar in Court

It was for the first time in 1954 that a bench of eight judges of the Supreme Court of India dealt with this contentious issue and ruled that the right to privacy is not protected by the Constitution.[92]

---

[90]Ministry of Home Affairs, Draft Bill on Right to Privacy, 29 September 2011. Available at: https://cis-india.org/internet-governance/draft-bill-on-right-to-privacy. Last accessed on 6 May 2020.

[91]Ministry of Electronics & Information Technology, Office Memo: Constitution of a Committee of Experts to deliberate on a data protection framework for India, 31 July 2017. Available at: https://meity.gov.in/writereaddata/files/meity_om_constitution_of_expert_committee_31072017.pdf. Last accessed on 6 May 2020.

[92]M.P. Sharma vs Satish Chandra, District Magistrate, Delhi, (1954) SCR 1077.

This ruling was further upheld in another judgement in the case of Kharak Singh vs the State of Uttar Pradesh[93] by a bench of six judges of the Supreme Court. However, post the Kharak Singh judgement, there were several other judgements delivered by the Supreme Court,[94] which held that the right to privacy was a fundamental right under Article 21 of the Constitution of India. These judgements were pronounced by benches with a lower strength and, therefore, doubt regarding the legal status of the right to privacy continued to prevail.

When petitions challenging Aadhaar were being heard by a bench of three judges of the Supreme Court, the Attorney General of India urged that in view of conflicting judicial decisions, the existence of a fundamental right to privacy is in doubt. The Supreme Court bench thereupon referred the matter to the Chief Justice of India, who constituted a bench of nine judges to decide upon the very limited question: whether right to privacy is a fundamental right protected by the Constitution or not.

This nine judges' bench[95] unanimously held that the right to privacy is protected as an intrinsic part of the right to life and personal liberty under Article 21 and as a part of the freedoms guaranteed by Part III of the Constitution. Thus, the right to privacy was declared a constitutionally protected fundamental right.

Thereafter, a five-judge bench heard all the Aadhaar-related petitions agitating the contentious issues relating to the Aadhaar architecture, the magnitude of protection accorded by the Aadhaar project to collection, storage and usage of biometric data, data protection, security and other key themes that have recurred since the beginning of the project.

Aadhaar was a bitterly argued case that led to the 'second-longest hearing in the history of the Supreme Court—spanning 38 working

---

[93]Supreme Court of India (From: Allahabad), Kharak Singh vs the State of Uttar Pradesh, 1964 SCR (1) 332, decided on 18 December 1962.
[94]For example, Gobind vs the State of Madhya Pradesh (1975), R. Rajagopal vs the State of Tamil Nadu (1994) and People's Union for Civil Liberties vs the Union of India (1997).
[95]Justice K.S. Puttaswamy (Retd) vs the Union of India, W.P. (Civil) No. 494 of 2012, decided on 24 August 2017.

days and nearly five months.[96]

After examining the evidence and arguments brought to its attention, the Court found that Aadhaar and its authentication services were so designed that they do not violate the right to privacy, nor could they enable mass surveillance.

The majority judgement found that the use of Aadhaar passes the 'triple test' laid down in the 'Privacy' judgement, under which there ought to be a law, a legitimate state interest and an element of proportionality in any law that seeks to abridge the right to privacy. On the basis of these considerations, the Court found the Aadhaar Act to be *intra vires*.[97]

The apex court also considered whether the Aadhaar Act defied the concept of limited government, good governance and constitutional trust. It found that Aadhaar not only gives the marginalized their unique identity, but also empowers them to avail of the fruits of government welfare schemes. The scheme ensures dignity to such individuals.

This happy outcome was no accident. It came about because the Aadhaar design follows the best practices for privacy and data protection, which itself was the consequence of careful debate and informed decisions at the inception of the project. Foresight backed by efforts have ensured that the Aadhaar design afforded all protections and yet met—and exceeded—the objectives for which it was created.

Social Media Profiling

Online social networks (OSNs) are pervasive, detailed and accessible repositories of personal information. Michal Kosinski of Stanford University wrote:

> [E]asily accessible digital records of behaviour—Facebook Likes—can be used to automatically and accurately predict a

---

[96]'Everything You Need to Know About the Aadhaar Case Before the SC Verdict,' The Wire, 26 September 2018. Available at: https://thewire.in/law/aadhaar-supreme-court-verdit. Last accessed on 1 June 2020.
[97]Justice K.S. Puttaswamy (Retd) vs the Union of India, W.P. (Civil) No. 494 of 2012, decided on 26 September 2018.

range of highly sensitive personal attributes including: sexual orientation, ethnicity, religious and political views, personality traits, intelligence, happiness, use of addictive substances, parental separation, age, and gender.[98]

Kosinski went on to show that judgements based on machine learning models can be more accurate than those of human beings[99] and that social media platforms other than Facebook are also useful in this respect.[100]

The worst fears indicated by these studies came true in the Cambridge Analytica story. In 2018, the company was accused of having 'harvested' 50 million Facebook users' data and used it without their consent for 'political advertising', besides engaging in other 'irregular' activities. The details shook public confidence in social media platforms and led to government investigations into their shenanigans, which revealed how personal information may be compromised, indeed traded, by these platforms or other parties who obtain access to their information, legitimately or otherwise.

Digital vulnerabilities arising from these networks have been modelled in terms of graph theory.[101,102] However, we should not

[98]Michal Kosinski, David Stillwell and Thore Graepel, 'Private Traits and Attributes Are Predictable from Digital Records of Human Behavior', *Proceedings of the National Academy of Sciences* 110.15 (2013): 5802–05. Available at: https://www.pnas.org/content/110/15/5802. Last accessed on 6 May 2020.

[99]Wu Youyou, Michal Kosinski and David Stillwell, 'Computer-based Personality Judgments Are More Accurate Than Those Made by Humans', *Proceedings of the National Academy of Sciences* 112.4 (2015): 1036–40. Available at: https://www.pnas.org/content/112/4/1036. Last accessed on 6 May 2020.

[100]D. Quercia, M. Kosinski, D. Stillwell and J. Crowcroft, 'Our Twitter Profiles, Our Selves: Predicting personality with Twitter', 2011 IEEE Third International Conference on Privacy, Security, Risk and Trust and 2011 IEEE Third International Conference on Social Computing, Boston, MA, 2011, pp. 180–85. Available at: https://ieeexplore.ieee.org/document/6113111. Last accessed on 6 May 2020.

[101]L.A. Cutillo, R. Molva and M. Onen, 'Analysis of Privacy in Online Social Networks from the Graph Theory Perspective', 2011 IEEE Global Telecommunications Conference—GLOBECOM 2011, Houston, TX, USA, 2011, pp. 1–5. Available at: https://ieeexplore.ieee.org/document/6133517. Last accessed on 6 May 2020.

[102]Bonneau, Joseph, et al. 'Eight Friends Are Enough: Social graph approximation via public listings', Proceedings of the Second ACM EuroSys Workshop on Social Network Systems, ACM, 2009. Available at: https://www.cl.cam.ac.uk/~fms27/papers/2009-BonneauAndStaETAL-friends.pdf. Last accessed on 6 May 2020.

conflate the vulnerabilities arising from these social networks with Aadhaar because in case of the latter, a graph can neither be constructed by UIDAI nor by the entity doing authentication. Unlike the social networks, where a large amount of information is public and visible, in the case of Aadhaar, the information collected is minimal and inaccessible to unrelated parties.

The law protects every piece of information involved in using Aadhaar, with the use itself restricted by the Supreme Court. On the other hand, the social networks hoard every picture, every line of text and every mouse click and sell it for the equivalent of the gross domestic product (GDP) of many countries. In this respect, Aadhaar's and OSN's are as different from each other as chalk and cheese.

The individuals who rant against Aadhaar on Facebook and Twitter don't grasp that their activities are being watched; their minds manipulated—as Kosinski found they could be and their rights, sold in the market for hundreds of billions of real dollars.

UIDAI has been strongly committed to data privacy since its very inception and has effectively embedded privacy throughout the programme's design—not as an afterthought.

The Aadhaar Act has articulated the PbD principles and given them a legal foundation. Furthermore, the Act and indeed the programme itself, are already in compliance with the major recommendations of the Srikrishna Committee. While the Aadhaar Act cannot replace an omnibus national privacy law, much needed in our country, it is certainly the first action our country has taken to legally ensure data privacy.

However, Aadhaar is only a technology tool. Technologies can be leveraged to deliver public good, but can also be misused if there are no checks and balances. We should remain vigilant in this respect because the right to privacy of people is vital for the functioning of a democracy like ours.

Chapter 6

CIVIL SOCIETY'S PUNCHING BAG

Building sustainable cities—and a sustainable future—will need open dialogue among all branches of national, regional and local government. It will need the engagement of all stakeholders— including the private sector and civil society, and especially the poor and marginalized.

—Ban Ki-moon, former Secretary-General, United Nations

UIDAI never underestimated civil society. From the outset, it proactively sought its participation as a strategic partner in designing the project. In perspective, it actually provided grist to the mill by arming civil society activists with information and cohesion to fight it. This was no matter of regret, as it nevertheless contributed to a robust design and enhanced sensitivity to the concerns of individuals as well as civil society.

In August 2009 itself, UIDAI recognized the importance of engaging with civil society organizations (CSOs) to improve the quality of its ideas in its Strategy Document. UIDAI's outreach programme was led by Raju Rajagopal, an engineering graduate of IIT Madras working with NGOs in India on poverty alleviation programmes, disaster management and communal harmony initiatives.

Raju prepared a Civil Society Outreach Plan in December 2009 to 'proactively engage a wide range of CSOs, especially those that speak for underserved and vulnerable communities'.

One of the first interactions with civil society representatives was organized on 30 and 31 October 2009 in Shimla at the Indian Institute of Advanced Study (IIAS). The list of participants included the who's who of civil society representatives from the academia and NGOs with a spread of sceptics and optimists. Issues that came up for discussion included privacy, exclusion, data misuse, information to the state, the need of certain communities to remain below the radar, possible corruption by the registrars, etc. It was a healthy debate.

There were half a dozen consultations following Shimla. These early meetings were energizing. The debates brought out points that would need careful examination and response in later design documentation and implementation plans.

A few members within UIDAI, even in the early interactions, sensed an ideological opposition from some CSOs. New ideas often do not fit within existing frameworks and, therefore, ideology-based opposition risks dousing the fire rather than igniting minds. We were prepared to be informed by ideology, but not prepared to let a closed mindset defeat fresh ideas.

Outreach, Opposition and Discord

Reaching out to prominent individuals in policy or social sectors, such as Renana Jhabvala, Arvind Kejriwal and others was a mixed experience. When Kejriwal did not attend a Data Standards Committee meeting, Raju called to enquire if he had been influenced by negative reporting. It turned out that Kejriwal had been busy with other commitments, but expressed full support for the project.

We held a major workshop with CSOs on 6 May 2010 in Vigyan Bhawan Annexe with over 40 representatives. I remember an interaction from the workshop in which a participant demanded, 'How could this project be started without consultation with civil society groups and their endorsement?' Being a bureaucrat, my instinctive reaction was that there is no procedure in the government to take clearance from civil society—and I said as much, which caused some consternation. Such outreach was indeed unusual, and for it to succeed, both sides needed to work towards the best interest of the society.

By the end of December 2009, there were several discussion groups and websites that came up regarding the UID. Dr Usha

Ramanathan, a privacy activist and a strong critic of the UID project, attacked it as a project of surveillance.[103] It is doubtful that based on working papers alone CSOs had fully grasped the technology, the manner in which it was proposed to be used and the safeguards to be built. But a position had been taken and the die was cast. It took many years and the Supreme Court to allay unfounded fears.

Consultations continued despite the discordant voices. I remember two in particular: in Guwahati on 15 February 2010 and in Pune on 2 March 2010. These discussions helped us think through the implications of Aadhaar's design and processes. While UIDAI adapted with each challenge and thus benefited from consultations, many constituents in civil society refused to come back for further analysis.

Raju summed up our outreach efforts concluding that while we were keen to partner with NGOs, several civil society leaders vehemently opposed the project. The project proceeded apace despite their opposition, but it unfortunately brought an abrupt end to UIDAI's civil society outreach efforts, precluding any effort to systematically respond to their concerns.

A group called 'Citizens against UID/Aadhaar' was started in the month of April 2010 which sought online support from citizens to oppose UID.[104] Reading the preamble of this site makes one feel as though if ever there was a single assault on the freedom of the people of this country, it is the UID project. Below is a mere sample of the kind of megalomania that this organization accused the UID project of being guilty of through their blog posts:

Kindly endorse the campaign against Unique Identification Number (UID), government of India's effort to give a number to every Indian citizen like they do to cattle before they are led to the slaughter house. It's kind of racial profiling which can be abused

---

[103]Usha Ramanathan, 'The Personal Is the Personal', *The Indian Express*, 6 January 2010. Available at: http://archive.indianexpress.com/news/the-personal-is-the-personal/563920/. Last accessed on 1 May 2020.

[104] Kindly Endorse: Citizens against UID/Aadhaar. Available at: https://tahaz.wordpress.com/2010/05/04/kindly-endorse-citizens-against-uid-aadhaar/. Last accessed on 11 May 2020.

in a country like India where they use the census data to butcher people in riots (remember the Gujarat pogrom). It's a violation of one's privacy, human rights and a threat to democracy.

There are numerous groups and websites whose only purpose seemingly is to condemn Aadhaar. Some provide a semblance of reasoning to support their arguments, while others are purely emotional appeals. One of the more popular of these contrarian organizations is Rethink Aadhaar.[105] Adherents of this organization were regularly pressed to tweet with such incendiary calls to action such as #DestroyTheAadhaar, etc. They even had a dedicated website (rethinkaadhaar.in) to their cause, with the following preamble:

> The Aadhaar project, to provide a unique number to all residents in India, was packaged as a welfare-enabling programme. It was sold as an initiative for greater inclusiveness in welfare, a tool against corruption, greater efficiency and so on.
>
> Six years down the line, the evidence shows the destruction of welfare programmes due to their Aadhaar linkage. The Aadhaar project also has a sinister side to it—it is a surveillance-enabling programme, which threatens privacy and democratic practice. Indian residents and citizens have been lulled into believing that privacy is the price they have to pay for better implementation of welfare programmes. In fact, Aadhaar project threatens both our welfare, and right to privacy. The Aadhaar project with its fundamental weaknesses and dangerous effects continues to grow.

These are like typical propaganda machines extracting and quoting examples from here and there to prove that Aadhaar has been an utter failure in all its good and declared objectives. These also assert that it has been clandestinely working for achieving the purpose for which the project was set up, viz. surveillance and destroying citizens' privacy.

However, let's set aside the rhetoric and look at the four broad categories of arguments against Aadhaar.

---

[105]Rethink Aadhaar. Available at: https://rethinkaadhaar.in/. Last accessed on 11 May 2020.

Aadhaar Is Ill-Conceived

One of the early criticisms was that UIDAI is 'doing its own thing' and ignoring the experiences of many countries on the issue of identity documents. There is the Social Security Number of the US and there are identity cards in various European countries. We should have learnt from those countries. Also, while the UK had abandoned its ID card project, we were implementing the same thing. Others advocated smartcards in the context of 'learning from others'. One article in Moneylife[106] published as early as July 2010, asserted that UIDAI will not be issuing a smartcard 'as was widely thought'.

UIDAI adopted the approach and strategy that was context-specific to India. Its focus was: to devise an identity system that ensured uniqueness and inclusivity. Biometrics came in because that was the only possible way to ensure uniqueness. Learning from us, some other countries are now considering the introduction of frugal, card-less online solutions to the problem of identity.

Biometrics Unreliable: It was claimed that biometric authentication is unreliable and a pipe dream. Since inception, there have been 40 billion authentications up to April 2020[107] and the number in some recent months has almost equalled the population of the country. There should be little doubt therefore that authentication does work for most of us, notwithstanding some exceptions. If authentication fails, in say 5 per cent cases, we need not abandon the technology, as it ensures delivery to the rest 95 per cent of people. The fact is, all technology fails sometimes: your mobile phone may not work when you have no signal!

Technology Untested: It was asserted that the technology adopted by UIDAI was untested. Some social scientists, such as Dr Ramakumar,

---

[106]Moneylife Digital Team, 'No Card, Only a Number Despite Rs 45,000 Crore Being Spent on the UID Project,' Moneylife, 7 July 2010. Available at: https://www.moneylife.in/article/no-card-only-a-number-despite-rs45000-crore-being-spent-on-the-uid-project/6920.html. Last accessed on 6 May 2020.
[107]Aadhaar Dashboard. Available at: https://uidai.gov.in/aadhaar_dashboard/. Last accessed on 6 May 2020.

opined that duplicates will be as high as 15 per cent, as the back-end deduplication technology will not be able to ensure uniqueness. Moreover, raising questions about scalability, the opinion was that while the deduplication technology could work in small samples, it simply would not work when applied to a billion people.

To be fair, Aadhaar was the first project in the world that had the audacity to go for this technology to deduplicate a billion. While the deduplication technology using fingerprints was relatively well understood, iris scan was not tested and tried at that scale.

Anyway, notwithstanding predictions by the prophets of doom, the technology worked as expected and has become a reference point for the world. Today, we have 1.25 billion Aadhaar numbers already allotted to residents of India, which nearly covers the entire population. These Aadhaar numbers have been issued after the required deduplication. While nobody can claim with certainty that a person with two Aadhaar numbers doesn't exist in the database, none can even remotely suggest that there are 200 million such individuals as was feared by Ramakumar!

Deduplication Not Needed: Some critics argued that Aadhaar was not required for deduplication at all because various agencies were already deduplicating their databases. Economist, social scientist and author, Reetika Khera[108] and Kieran Clarke,[109] argued that deduplication of liquefied petroleum gas (LPG) connections was being done through digitization of records and list-based methods.

However, there are two points to consider that militate against such deduplication. Firstly, demographic deduplication by combining data from multiple databases is not effective because people game

---

[108]Reetika Khera, 'The UID Project and Welfare Schemes,' *Economic & Political Weekly*, Vol. 46 (9), 26 February 2011. Available at: https://www.epw.in/journal/2011/09/perspectives/uid-project-and-welfare-schemes.html. Last accessed on 6 May 2020.

[109]Kieran Clarke, 'More Ghost Savings: Understanding the fiscal impact of India's direct transfer program—update,' International Institute for Sustainable Development (IISD), February 2016. Available at: https://www.iisd.org/library/more-ghost-savings-understanding-fiscal-impact-india-direct-transfer-program-update. Last accessed on 6 May 2020.

the system by writing their names in multiple ways. Secondly, such cleansing is typically drive-based. You conduct a drive to clean the data, such as cancellation of bogus and duplicate ration cards, but after some time, the number swells again due to new ration cards being made by interested parties!

While Khera conceded that biometrics can be useful for deduplication, she argued that the purpose can just as well be served by a local, rather than an online, centralized biometric database.[110] She gives the example of Andhra Pradesh, where local biometric databases were being successfully used for deduplication before Aadhaar.

The Andhra example of local biometrics having been successfully used for deduplication is factually not correct. While the Government of Andhra Pradesh had collected the iris images of 55 million people, it never completed the task of deduplication.

No Savings: Some civil society activists contended that the project is not justified on the basis of actual or expected benefits or savings. Extrapolating from a study by the London School of Economics of the UK's own ID project, *Frontline*'s Praful Bidwai[111] implied that the cost of the project may 'exceed ₹2 lakh crore'. The actual amount spent by UIDAI came to less than ₹10,000 crore (₹100 billion) or less than ₹100 per Aadhaar due to its careful design. Compare this to the UK's ID card cost estimated at 'roughly £170 per card and passport.'[112]

Many prominent persons dismissed the cost-benefit analysis done by the National Institute of Public Finance and Policy (NIPFP), which stated:

---

[110]Reetika Khera, 'Debate: Five Aadhaar myths that don't stand up to scrutiny,' The Wire, 23 March 2016. Available at: https://thewire.in/25578/rebooting-the-aadhaar-debate/. Last accessed on 6 May 2020.

[111]Praful Bidwai, 'Questionable Link,' *Frontline*, Vol. 27 (12), 18 June 2010. Available at: https://frontline.thehindu.com/columns/article30180769.ece. Last accessed on 6 May 2020.

[112]'The Identity Project, ID Cards—UK's high-tech scheme is high risk,' London School of Economics study, 27 June 2005. Available at: http://www.lse.ac.uk/management/research/identityproject/PR1.htm. Last accessed on 6 May 2020.

Even after taking all costs into account, and making modest assumptions about leakages of about 7–12 per cent of the value of the transfer/subsidy, we find that the Aadhaar project would yield an internal rate of return in real terms of 52.85 per cent to the government.[113]

While making its projections, NIPFP had taken the cost of the project to be a modest sum of ₹14,000 crore (₹140 billion). UIDAI, spent 30 per cent less.

Other studies and reports have since concluded that Aadhaar has enabled huge savings through the elimination of fakes and duplicates in various databases, including LPG, PDS, scholarships, National Social Assistance Programme (NSAP), MGNREGA and midday meals. Improved service delivery has ensured that benefits reached the intended beneficiaries (through Direct Benefit Transfer [DBT], PDS, AEPS) and the cost of service delivery through DBT, Aadhaar Pay, eKYC and eSign has been reduced.

A prominent programme where cost savings were achieved through DBT of subsidy is PAHAL (Pratyaksh Hanstantrit Labh) in LPG distribution. Clarke,[114] Venkatanarayanan[115] and Khera[116] in separate studies claimed that either the savings were exaggerated because of wrong estimates of underlying numbers or due to misattribution of savings to Aadhaar instead of other causes, such as drop in gas prices. It was, however, left to the Comptroller and Auditor General (CAG) of India to point out that Aadhaar had

---

[113]R.S. Sharma, 'Aadhar Is Transparent and Accountable,' *The Hindu*, 20 March 2013. Available at: https://www.thehindu.com/opinion/op-ed/aadhar-is-transparent-and-accountable/article4526474.ece. Last accessed on 6 May 2020.

[114]Kieran Clarke, 'More Ghost Savings: Understanding the fiscal impact of India's direct transfer program—update,' International Institute for Sustainable Development (IISD), February 2016. Available at: http://www.iisd.org/sites/default/files/publications/more-ghost-savings-india-direct-transfer-program-policy-brief.pdf. Last accessed on 6 May 2020.

[115]Anand Venkatanarayanan, 'Government's Claims of Aadhaar Savings for the LPG Scheme Are Overstated,' Medianama, 2017. Available at: https://www.medianama.com/2017/06/223-aadhaar-lpg-scheme/. Last accessed on 6 May 2020.

[116]Reetika Khera, 'Debate: Five Aadhaar myths that don't stand up to scrutiny,' The Wire, 23 March 2016. Available at: https://thewire.in/government/rebooting-the-aadhaar-debate. Last accessed on 6 May 2020.

not been fully seeded in the list of PAHAL beneficiaries[117] and, therefore, the savings in these studies were underestimated.

The aggregate savings in all government benefit programmes, prevention of tax evasion by linking PAN numbers and bank accounts, and the efficiency gains within the government would be substantial compared to the total cost of Aadhaar, which may be of the order of rounding errors in these estimates.

Efficiency improvements in the economy accrue outside the government too. For instance, mobile companies can enable eKYC, thereby reducing the cost of customer acquisition and lowering the cost of preserving a trainload of paper records. Banks, cab aggregators (such as Ola, Uber), e-commerce portals (such as Amazon, Flipkart) and almost anyone in the digital economy could benefit from a verifiable online ID system.

Database Insecure: Sunil Abraham, technology policy analyst from the research organization The Centre for Internet & Society, argued that the Aadhaar database is insecure by its very design.[118] He points to two characteristics of the database, viz. its centralized design and the backdoors in the database created for the government as major security risks. But this is a classic strawman fallacy.

There are no backdoors for the government into the system. None were requested nor were any provided. Moreover, biometric data is used only for deduplication and authentication. It is safely stored—with a high level of encryption—in CIDR at UIDAI. Further, UIDAI doesn't collect any data other than the basic information already printed on the Aadhaar card and contained in the authentication logs.

Fraud Promotion: Abraham also argues that the use of biometrics for authentication is dangerous and smartcards with digital

---

[117]Ministry of Petroleum & Natural Gas, Report of CAG on Implementation of PAHAL (DBTL) Scheme for the period ended 31 March 2016 (Report no. 25 of 2016). Available at: https://cag.gov.in/sites/default/files/audit_report_files/Union_Commercial_Compliance_Full_Report_25_2016_English.pdf. Last accessed on 6 May 2020.
[118]Sunil Abraham, 'Surveillance Project', *Frontline*, 15 April 2016. Available at: http://www.frontline.in/cover-story/surveillance-project/article8408866.ece. Last accessed on 6 May 2020.

signatures are a far safer option. He states that biometrics can be spoofed to create fake Aadhaar IDs that pass manual scrutiny for which there are free tools available online, such as SFinGe (Synthetic Fingerprint Generator) that allow you to create fake biometrics. With a little bit of clever programming, countless number of ghosts can be created, which will easily clear the manual adjudication process that UIDAI claims will ensure 'no one is denied an Aadhaar number because of a biometric false positive'.

Journalist Sahil Makkar, in an interview with Sunil Abraham, makes a similar point regarding iris data.[119] He states that 'it is easy to capture the iris data of any individual with the use of next-generation cameras. Imagine a situation where the police is secretly capturing the iris data of protesters and then identifying them through their biometric records.'

Bharadwaj[120] further cites examples of Axis Bank, Suvidhaa Infoserve and eMudhra to demonstrate that biometrics-based fraud is not only possible, but has already taken place.

Now, frauds can take place in any system and there is a law in place to take care of such attempts. The efforts of all technology-based systems should be to reduce the possibility of such frauds being committed. For example, cancellable biometrics and fraud detection engines at the authentication back end look for patterns that suggest fraudulent use, similar to what is done by credit card companies such as Visa and Mastercard.

Aadhaar Is Illegal

Critics claim that Aadhaar is not backed by law, therefore it is illegal. The question is: could UIDAI start the project without a law in place?

---

[119]Sahil Makkar, 'Aadhaar Is Actually Surveillance Tech: Sunil Abraham', *Business Standard*, 12 March 2016. Available at: http://www.business-standard.com/article/opinion/aadhaar-is-actually-surveillance-tech-sunil-abraham-116031200790_1.html. Last accessed on May 2020.

[120]Kritika Bharadwaj, 'Explainer: Aadhaar is vulnerable to identity theft because of its design and the way it is used', Scroll.in, 2 April 2017. Available at: https://scroll.in/article/833230/explainer-aadhaar-is-vulnerable-to-identity-theft-because-of-its-design-and-the-way-it-is-used. Last accessed on 6 May 2020.

Legitimacy: We posed this issue to the Planning Commission, which was our anchor department. It sought the view of the Attorney General, who opined that the Authority functioning under the executive notification dated 28 January 2009, had valid authority. There was nothing in law or otherwise that could prevent the Authority from functioning under Executive Authorization. The power of the Executive is clear and there is no question of circumventing Parliament or the Executive becoming a substitute of Parliament. All the expenditure which is being incurred is sanctioned by Parliament in accordance with the financial procedure set forth in the Constitution.

I must admit however, that the absence of a law became a debilitating factor for Aadhaar. In a sense, it did lack full legitimacy. While UIDAI dealt with the issues raised by the Standing Committee on Finance and the same was submitted to the Union Cabinet, no steps were taken at that time to pursue the line to pass the law. It came much later in 2016.

Function Creep: Ramanathan argued that the scope of Aadhaar's functions has been expanded time and again beyond its original intent of offering an identity proof to the poor, to becoming a mandatory ID so much so that not having it can be construed as a crime. Her arguments broadly run as follows:

In the case of the UID project, it was promoted as providing the poor with an identity. Then, it was about deduplicating the entire population, so that each person would have one unique number by which they could be identified. Then it was to get rid of 'ghosts' and 'duplicates' in welfare systems and to prevent 'leakages'. Then, from denial of entitlements—if a person is not enrolled or does not seed their number—it has reached a point where not having a UID number will force you to commit an offence. Why? Because you want to pay your taxes, but cannot, as the government will refuse to accept tax payment from you if you do not give your UID number. Then the government will cancel your PAN card and levy a penalty because you do not have a PAN card.[121]

---

[121]Usha Ramanathan, 'The Function Creep That Is Aadhaar', The Wire, 25 April 2017.

The scope creep described above is conjured up by glib use of language. 'Unique' is in UIDAI's name itself, so 'deduplicating the entire population' is not an afterthought. The purpose of deduplication is to get rid of the duplicates in welfare systems, not an exercise in amusement.

And what about the fantastic claim of forcing you to commit an offence? If the government asks for evidence of your property details, which you are loathe to provide, does it amount to forcing you to not pay taxes? The government has the responsibility to seek information necessary to prevent tax frauds, and proof of identity is one such requirement. Other identity documents have been used before Aadhaar, but did not work because 'benami'[122] accounts and transactions could help in the cover-up.

UIDAI, in its Strategy Document had specified that Aadhaar is voluntary. However, it was also mentioned that while Aadhaar is voluntary from UIDAI's perspective, it may be made compulsory if a domain using Aadhaar so requires. For example, it may be made compulsory by the Department of Food and Civil Supplies of a state for accessing subsidized ration.

When the government hands out a subsidy, it has the right— rather a responsibility—to ensure that the subsidy reaches the intended recipient and no one else. For that, it may seek a proof of identity that uniquely identifies the person. Similarly, when people try to evade taxes, the government has the right to take measures against the evasion. Aadhaar makes it difficult to commit frauds by linking accounts and transactions to a unique, real person. The Supreme Court settled this debate and upheld the government's right to ask for Aadhaar in specific cases.

Surveillance: Several commentators such as Drèze,[123] Jyoti

Available at: https://thewire.in/128039/aadhaar-function-creep-uid/. Last accessed on 6 May 2020.
[122]'Benami' is essentially an Indian origin word which means holding in someone else's or a fictitious name to cover up the identity of the beneficial owner.
[123]Jean Drèze, 'Unique Facility, or Recipe for Trouble?' *The Hindu*, 25 November 2010. Available at: http://www.thehindu.com/opinion/op-ed/Unique-facility-or-recipe-for-trouble/article15714630.ece. Last accessed on 7 May 2020.

Panday,[124] Ramanathan,[125] Khera[126] and Shyam Diwan[127] have stated that Aadhaar is a perfect tool for surveillance pretending to be a development programme, which will have a chilling effect on free speech and individual liberty.

This is an unwarranted concern from many viewpoints. Firstly, there are many devices in existence that can track you better than Aadhaar. There are CCTV cameras everywhere. Your smartphone emits a huge amount of data relating to your movements, call records and other online activities and governments have legal authority for surveillance and tracking under various laws, such as the Indian Telegraph Act, 1885.

In a connected world, there is a huge amount of data about all of us that is publicly available. It's discoverable through a simple Google search. Then, there is data available in relationship graphs derived from social media platforms. And yet more data that is not searchable, but exists with large corporates such Google, Amazon, Facebook or Apple.

About a typical Indian resident, UIDAI has the least amount of data compared to any of these actors. The data it accumulates in use is only authentication logs, which themselves don't contain anything except what is required to establish an authentication request was made and answered.

Apprehensions about the creation of a surveillance state have been examined by the Supreme Court that held that Aadhaar doesn't

[124]Jyoti Panday, 'Aadhaar: Ushering in a commercialized era of surveillance in India,' Electronic Frontier Foundation, 1 June 2017. Available at: https://www.eff.org/deeplinks/2017/05/aadhaar-ushering-commercialized-era-surveillance-india. Last accessed on 7 May 2020.

[125]Usha Ramanathan, 'Coercion and Silence Are Integral Parts of the Aadhaar Project,' The Wire, 16 May 2017. Available at: https://thewire.in/136102/coercion-aadhaar-project-ushar/. Last accessed on 7 May 2020.

[126]Reetika Khera, 'The Different Ways in Which Aadhaar Infringes on Privacy,' The Wire, 19 July 2017. Available at: https://thewire.in/159092/privacy-aadhaar-supreme-court/. Last accessed on 7 May 2020.

[127]Shyam Diwan, 'The Aadhaar Trap: Why you should be really, really worried,' First Post, 21 December 2014. Available at: https://www.firstpost.com/business/economy/you-should-be-worried-with-aadhaar-you-are-at-govts-mercy-1315823.html. Last accessed on 7 May 2020.

lead to the creation of a surveillance state.[128] Do not fear using your Aadhaar in any situation where you do not fear giving out your name.[129]

Aadhaar Is Unnecessary

UIDAI's starting assertion in its Strategy Paper itself was that an 'inability to prove identity is one of the biggest barriers preventing the poor from accessing benefits and subsidies. Till date, there remains no nationally accepted, verified identity number that both residents and agencies can use with ease and confidence.'[130]

Identity Is Not a Problem: CSOs, however, claimed that it is not the lack of identity that denies access to resources for almost three-fourth of the population in India, but rather other problems that plagued the implementation of various programmes.

In an email, Dr Tushar Kanjilal[131] stated: 'Political parties have ensured that people are given either a ration card or the voter ID. So, lack of ID documents is not the problem. Whether they actually get the rations is the real problem!' It was hence claimed that the real problem was misclassification of beneficiaries as a result of a flawed BPL census, not of identity fraud or duplicate cards.

According to Khera, only 0.03 per cent of those who have Aadhaar used the 'introducer' provision to get enrolled. The remaining enrolled using some other proof of identity and proof of address. She argues that the lack of ID documents was not

---

[128]Writ Petition (Civil) No. 494 of 2012.

[129]R.S. Sharma, 'Have No Fear: Aadhaar is linked to logic,' *The Economic Times*, 13 February 2018. Available at: https://economictimes.indiatimes.com/news/economy/policy/view-have-no-fear-aadhaar-is-linked-to-logic/articleshow/62890964.cms. Last accessed on 7 May 2020.

[130]'UIDAI Strategy Overview: Creating a unique identity number for every resident in India,' page 1. Available at: https://www.prsindia.org/sites/default/files/bill_files/UIDAI_STRATEGY_OVERVIEW.pdf. Last accessed on 7 May 2020.

[131]Head of the Tagore Society for Rural Development (TSRD), Kolkata, widely known for his fight to save the Sundarbans, and featured in Amitav Ghosh's *The Hungry Tide*. TSRD works with the most marginalized communities in West Bengal as well as parts of Jharkhand and Odisha.

a problem, which UIDAI was trying to solve in the first place. Khera also argued that Aadhaar does not have much role to play in reducing leakage and corruption in MGNREGA or PDS.

The fact is that ID in the UIDAI documents refers to a 'nationally accepted verified identity document, which both the government and the private sector can use for service delivery'. For example, the ration card, typically a family document, can no longer be used as an ID document to open a bank account.

The assertion that miniscule numbers used the introducer system for Aadhaar enrolment is far from the truth. As per statistics with UIDAI, a substantial number—more than 30 per cent—used either the family document (ration cards) or the letter signed by someone as a proof of identity and proof of address.[132] None of these documents could be used as a general-purpose ID document. This data disproves the assertion that most people had identity documents.

UIDAI never claimed to solve all the problems afflicting beneficiary programmes. It is merely removing the disability of people in accessing the formal system due to lack of an ID. And where allocation of resources is inadequate for the purpose, it can also prevent some of the benefits from reaching the wrong hands.

Aadhaar has been a boon for those without ID documents. For example, when we carried out a special enrolment camp at Yamuna Pushta, an illegal colony in Delhi, we found that most of the people living there for several years did not have any ID papers. They could not open a bank account and had to keep their money as cash in their hutments. Police would harass them for not having any ID papers and their money would get stolen. Aadhaar helped them. Were the residents of Yamuna Pushta, and I'm sure many others like them, not visible to the civil society groups?

Ghost Beneficiaries—Exaggerated Issue: One reason for starting the 'unique' ID programme was the existence of multiple and

---

[132]Proof of Identity (POI) and Proof of Address (POA) Documents used in Aadhaar Enrolment, Available at: https://drive.google.com/file/d/0BzhU1mYrqTDuYWlzMWNsc U5NY1E/view. Last accessed on 6 May 2020.

ghost IDs in beneficiary programmes and consequent leakages. However, some CSOs contended that the problem of duplicate and fake beneficiaries was exaggerated. Their numbers were small. Further, there was leakage even before the ration reached the shops. Moreover, Aadhaar could not help with identifying or removing inclusion errors, where a person was wrongly included as a beneficiary.

UIDAI never asserted that it could solve the problem of wrong inclusions in government programmes. However, with Aadhaar seeded in separate benefit delivery databases, the government could cross-verify the beneficiaries of these programmes, where justified by need and supported by law. For example, eligibility conditions debar a person from one programme, if he or she is the beneficiary of another one: you are not eligible for subsidized kerosene oil while also availing subsidized LPG. UID could eliminate such cases.

It is true that duplicates have been weeded out in the past in many states. But they come back soon after. Aadhaar-based deduplication not only solves the problem, but ensures that a duplicate ration card cannot be created again. Biometric authentication at the point of service delivery acts as an independent proof of presence and delivery. Not only that, as the PDS transaction is online, it can confirm who got the ration at what time and how much. This provides a real-time inventory status of goods at the fair price shop (FPS). It solves the problem of the FPS dealer fraudulently showing issue of the full quantity of ration to card holders for diversion into the black market.

Further, biometric authentication can enable portability as now both the identity and entitlements are online and you can get ration from any shop. Considering the large number of migrant workers in our country, a number of states have introduced intra-state PDS portability.[133] One can see real-time lifting of foodgrains from the shops on state web portals. Aadhaar-enabled PDS (AePDS)

---

[133]News Express Service, 'Task Force on Migration: Panel seeks PDS transfer within states, better data collection,' *The Indian Express*, 10 September 2016. Available at: https://indianexpress.com/article/india/india-news-india/task-force-on-migration-workers-jobs-sanitation-pds-within-states-3023186/. Last accessed on 7 May 2020.

introduced in Andhra Pradesh is an example of transparency. The Jharkhand portal of PDS called Aahar[134] also provides real-time lifting, stock and other details of PDS's working in the state.

Aadhaar Is Evil

Proponents in this category assert that due to biometric authentication at the point of service delivery, Aadhaar shall result in exclusion of beneficiaries from social safety programmes such as MGNREGA and PDS. Two reasons that are offered to support this exclusion argument was that many may not be able to get Aadhaar or authentication may fail at the point of service delivery.

Exclusion of Beneficiaries: When the Andhra Pradesh government integrated their PDS with Aadhaar, it found that nearly 20 per cent of the beneficiaries stopped buying rations. In a follow-up study,[135] out of the 790 beneficiaries interviewed, 490 reported exclusion. The reasons were fingerprint mismatch (290), Aadhaar number mismatch (93) and failure of electronic point of sale (EPOS) devices (17).

Mismatches may result either when the Aadhaar number does not match with the ration number, or when beneficiaries' names do not tally with the Aadhaar record. Other reasons stated for exclusion were that the ration shop dealer does not know how to use the device or discourages beneficiaries from using the EPOS machine.

Similarly, in Rajasthan, Anumeha Yadav[136] notes that Aadhaar was made mandatory for receiving rations under the PDS. However, beneficiaries from several villages did not receive their

---

[134]Aahar: PDS Portal of Govt. of Jharkhand. Available at: https://aahar.jharkhand.gov.in. Last accessed on 6 May 2020.

[135]'FP Shops Left over Beneficiaries Report,' Society for Social Audit, Accountability and Transparency, 2015.

[136]Anumeha Yadav, 'In Rajasthan, There Is "Unrest at the Ration Shop" Because of Error-Ridden Aadhaar,' Scroll.in, 2 April 2016. Available at: https://scroll.in/article/805909/in-rajasthan-there-is-unrest-at-the-ration-shop-because-of-error-ridden-aadhaar. Last accessed on 7 May 2020.

entitlements due to authentication failures, errors in seeding, fingerprint mismatches, name mismatches and errors in recording or entering demographic information during enrolment. Further, some beneficiaries were permanently locked out of the system due to repeated authentication failures. This problem is exacerbated even further due to a lack of easy access to grievance redress for Aadhaar-related issues.

There were also other reasons for exclusion, such as in Chitradurga, Karnataka, in 2014–15, when district officials deleted the job cards of many beneficiaries to show '100 per cent Aadhaar seeding', which led to their MGNREGA wages being held up.

The first issue of coverage now stands addressed, as almost the entire population has been covered. UIDAI has ensured enrolment facilities to those who were left out.

Let us now look at the other reasons for exclusions.

Firstly, UIDAI has all along advocated that the service delivery organizations using Aadhaar authentication must account for failure of authentication in a small percentage of the population. This can happen for a variety of reasons and it is, therefore, imperative that the service delivery organizations put in place an exception-handling mechanism. Technology never works 100 per cent of the time, more so in our country where power may go out, connectivity may fail or the device itself may not perform as expected.

We must incorporate a mechanism to handle such cases. After all, it is not a vending machine delivering the ration. There is a human being at hand who can deal with such cases (usually less than 2-3 per cent) through manual verification. Total reliance on technology systems is untenable and must be avoided.

As a matter of policy, the service delivery agencies must make a rule that if the authentication fails after a given number of attempts, the person should be given the ration, after manually entering the details in the delivery system.

It is to be noted that there are many exogenous factors why authentication may fail. For instance, if there is an error in seeding the Aadhaar number in the PDS entitlement records, authentication would fail. It is akin to a cheque being dishonoured because of a wrong account number or mismatch of signature. Indeed, when the error is deliberate, it is called a fraud and the system is expected to result in an authentication failure.

It isn't difficult to see why someone may deliberately insert an incorrect Aadhaar number and make only feeble attempts to rectify the error. The FPS owner may prefer to let authentications fail if it allows him to divert the supplies to the black market or the government servants and village-level workers may be complicit in these shenanigans when they have a share in the ill-gotten gains. Clearly, in these cases, UIDAI is only the proverbial fall guy.

The way to neutralize such actors would be to provide the direct beneficiary a reason to opt for biometric authentication. For example, if a one-time pay out is given to the beneficiary for successful biometric authentication, she may insist upon it. From the other end, the government too must push for authentication using administrative measures. The two together may be able to squeeze out the corrupt middlemen.

From the technology perspective, we should deploy devices with both iris and fingerprint-based authentication, which will reduce authentication failures to less than a percentage point. This should be in addition to the exception-handling systems as stated above.

As the PDS data seeding is being cleaned up, instances of authentication failure are reducing.

Beginning of the End of Welfare Schemes: Many CSOs saw the introduction of Aadhaar as the first step in dismantling welfare schemes such as MGNREGA and PDS and replacing India's public services with conditional cash transfers (CCTs) or DBTs. Drèze criticizes this transition for favouring businesses instead of achieving better development outcomes. It was moreover claimed that CCT/DBT allows a state to abscond from some of its responsibilities of improving the lives of its citizens from conditions of malnourishment, unemployability and access to education and social services.

Is this an argument against DBT or Aadhaar? Inefficiencies and corruption in PDSs result in costly, untimely and erratic deliveries of benefits. Seeds reaching the farmers after the sowing time, books reaching the students much after the academic session has commenced and foodgrains not reaching the ration shops are some examples. Cash transfers can make it better. Cash or kind, technology can help in improving the efficiency and targeted delivery of these programmes.

While CSOs accepted that DBT by itself is a good way to curb corruption in delivering social welfare, it is possible and—indeed better—to have a system without Aadhaar. Kapur and Khera argue that:

> What the government had in mind when it spoke of DBT is actually 'Aadhaar-enabled DBT' to transfer cash or subsidy. That requires three things: a modern banking sector to which beneficiaries have access; Aadhaar number for all beneficiaries and seeding bank accounts with Aadhaar. Only when all three are in place, can Aadhaar-enabled DBTs proceed. Eight months after DBT was launched, 56 per cent of beneficiaries had a bank account; 25 per cent had an account and Aadhaar; but only 9.6 per cent had all three. This caused havoc. When Aadhaar was made compulsory for cash transfers, it led to exclusion on a massive scale. For example, many elderly people stopped getting their pension in Jharkhand.[137]

Many have argued that bringing in Aadhaar for DBT unnecessarily complicates things without adding any value. DBT can be done without Aadhaar. It is natural to ask once you have the person and her bank account details, why should you need another variable such as Aadhaar? Using a bank account number for DBT is error-prone and can be misused because account numbers don't solve the problem of duplicates. Using Aadhaar cleans up the system and one does not have to seed bank account details in every domain.[138]

---

[137]Manavi Kapur, 'Aadhaar-Enabled DBT Is More Demanding Than DBT: Reetika Khera,' *Business Standard*, 29 June 2014. Available at: http://www.business-standard.com/article/current-affairs/aadhaar-enabled-dbt-is-more-demanding-than-dbt-reetika-khera-114062800476_1.html. Last accessed on 6 May 2020
[138]R.S. Sharma, 'Aadhaar Bill: Click out the missing link,' *The Economic Times*,

Surveillance Project: CSOs also asserted that Aadhaar was a surveillance project that would help US companies as the data reaches the US/CIA.[139] It was claimed that Aadhaar was an assault on human rights and being bulldozed at the behest of US biometric technology companies that manifestly worked with intelligence agencies and would ensure profiling of minorities and neutralize political resistance in India. Also, UIDAI as the custodian of Aadhaar data, had already signed contracts with foreign firms giving them 'full access' to unencrypted personal and classified data of residents, which they were allowed to store for seven years.[140] Moreover, the government was coercing citizens to enroll in order to sell the data to corporate interests.

The problem is that these authors—without knowledge and without the obligation to offer evidence—made assertions to mislead by insinuation and innuendo. There is nothing new about the government working with foreign companies and nothing against it, if adequate safeguards are put in place. The entire resident data is kept encrypted, sharded (rows of a database are held separately) and vertically partitioned. For instance, the biometric images are stored separately from demographic information.

BSPs, who provide the biometric deduplication service, work competitively on a plug-and-play basis. They only have read access to the biometrics and no idea whose biometrics these are. They just have a reference ID, which they use to return the deduplication results. The architecture has been planned to be on minimal knowledge basis. As the BSPs have nothing to do with Aadhaar number generation or any other aspect of the back-end process, their only job is to read the biometric features, do the deduplication

---

5 March 2016. Available at: https://economictimes.indiatimes.com/blogs/et-commentary/aadhaar-bill-click-out-the-missing-link/. Last accessed on 7 May 2020

[139]Gopal Krishna, 'Boycott UID/Aadhaar Number', Countercurrents.org, 27 September 2011. Available at: https://www.countercurrents.org/krishna270911.htm. Last accessed on 7 May 2020

[140]Sandhya Jain, 'Aadhaar, Data Security and Breach of Privacy', 5 September 2017, citing an RTI application filed by Col Matthew Thomas. Available at: https://www.dailypioneer.com/2017/columnists/aadhaar-data-security-and-breach-of-privacy.html. Last accessed on 11 May 2020.

process from the gallery of images and return the results. If there is a hit, i.e., the biometrics matched with someone in the gallery, they return the reference ID of the matched biometrics, else they return with the information that there was no hit. Other back-end processes do the rest of the job.

BSPs do not have any control over the data they will receive. In fact, their share is decided by a formula based on their accuracy, speed and requirements of hardware. BSPs can be suspended if they are not able to maintain the required standards and UIDAI has, indeed, changed BSPs without causing any disruption.

In fact, the UIDAI technology architecture is one of the best-designed architectures that ensures accuracy, speed, cost optimization, no vendor lock-in and security of the data.

Not only have activists criticized Aadhaar and its design, they have also criticized UIDAI and other government agencies. The key criticisms levied against the Authority relates to data leaks, lack of transparency and overlooking data breaches.

Ramanathan[141] contends that Aadhaar-based data is likely to leak personal and sensitive information about citizens, and gives the example of Jharkhand, Kerala and Bihar.

> In Jharkhand, a 'glitch' resulted in the exposure of the personally identifiable information—including names, addresses, UID numbers and bank account details of those drawing old age pensions. That is about 1.6 million people, of whom 1.4 million have seeded their UID numbers. In Kerala, sensitive personal information of 34 million people has been leaked in what is described as 'one of the biggest data breaches in the world.' In Bihar, the name, address, bank account number, bank address, the ID of the parents and the UID number of students getting a post-matriculation scholarship was displayed on the net for anyone to see.

She argues that it does not take nefarious intent for such leaks

---

[141]Usha Ramanathan, 'The Function Creep That Is Aadhaar', The Wire, 25 April 2017. Available at: https://thewire.in/128039/aadhaar-function-creep-uid/. Last accessed on 6 May 2020.

to occur, but rather just carelessness or negligence on the part of agencies in possession of this data.

Yadav[142] states that UIDAI is under no obligation to inform affected citizens in case their data is compromised and the Aadhaar Act does not provide any right to a citizen to seek legal remedy in case their data is breached.

There is a law in the country about how data breaches are to be dealt with. There is another law about how Aadhaar may be used by the government and what rights an individual has against possible misuse. Also, the examples of data leaks splashed in the media do not qualify to be called data leaks as has been explained in the next chapter of this book. That said, it is important to build a good data-protection framework and set of practices to go with it that would evolve over time, as they must. But this is no argument against the use of a technology that is effective and has passed the constitutional tests in the Supreme Court. Regarding UIDAI's lack of transparency and oversight with respect to breaches, The Wire[143] claimed that the Authority had refused to give information on fake and duplicate Aadhaar cards by citing a national security exception under Section 8(a) of the Right to Information (RTI) Act.

Yadav notes that contrary to international best practices, Aadhaar lacks proactive disclosure rules, and keeps citizens in the dark regarding data breaches and security incidents. She also cites the example of writer Sameer Kochhar against whom UIDAI filed a criminal complaint for demonstrating how the Aadhaar database could be hacked.[144] The Wire[145] cites another example where a

---

[142]Anumeha Yadav, 'Under the Right to Information Law, Aadhaar Data Breaches Will Remain a State Secret,' Scroll.in, 5 March 2017. Available at: https://scroll.in/article/830589/under-the-right-to-information-law-aadhaar-data-breaches-will-remain-a-state-secret. Accessed on 6 May 2020.

[143]The Wire Staff, 'UIDAI Cites "National Security" To Block RTI Query on Fake and Duplicate Aadhaar Cards,' The Wire, 12 June 2017. Available at: https://thewire.in/government/uidai-invokes-national-security-reject-rti-query-fake-duplicate-aadhaar-cards. Last accessed on 7 May 2020.

[144]The Wire Staff, 'UIDAI Files FIR against the Tribune, Reporter Over Aadhaar Breach Story: Report,' The Wire, 7 January 2018. Available at: https://thewire.in/tech/uidai-files-fir-tribune-reporter-aadhaar-breach-story-report. Last accessed on 7 May 2020.

[145]ibid.

criminal complaint was filed against TV journalist Debayan Roy for airing a segment in which he explained how he was able to get two Aadhaar IDs using the same set of biometrics.

Ethical hacking, as referred to above, has a code of practice. The security flaws that are discovered by ethical hackers are disclosed to the organization whose security system is being tested. Disclosure of private information or the method of public attack remains a crime under certain circumstances and is open to action under the law.

Car racing is a sport. Racing on the streets is illegal. One cannot be confused with the other in discussions. Similarly, media sensationalism and risky or injurious behaviour by the so-called hackers is not ethical hacking and has no legitimacy. However, it is for UIDAI to deal with these and for lawyers, lawmakers and courts to help in the development of the law of the land.

Who Is the Final Arbiter?

The most ancient of texts, the Rig Veda, lays down how any public policy must be judged with the words: '*bahujana sukhaya bahujana hitaya ch*' meaning 'for the happiness of the many, for the welfare of the many'. None can claim to know what constitutes the happiness of the people. The people alone can proffer the definitive answer.

It's been a decade since UIDAI was constituted. By now, more than 1.25 billion Indian residents have voluntarily obtained their unique identification or UID number. There was never an incident of serious protest by these people or the need to use force by any government agency, the misgivings of some sections of civil society notwithstanding. Indians have not only got their Aadhaar number, but are using it too, as evidenced by the increasing number of authentications every month.

A recent survey of 1,67,000 Indian residents by Dalberg, the global consulting firm,[146] (supported by Omidyar Network) has found that 92 per cent people are satisfied with Aadhaar, 90 per

---

[146]Swetha Totapally, Petra Sonderegger, Priti Rao, Jasper Gosselt and Gaurav Gupta, *State of Aadhaar Report 2019*. Dalberg, 2019. Also, visit: https://dalberg.com/our-ideas/dalberg-host-state-aadhaar-initiative/.

cent of them trust their data is safe and 80 per cent feel it has made PDS rations, rural employment guarantee schemes such as MGNREGA and other social programmes more reliable.

It is truly heart-warming that nearly 50 per cent of the people used Aadhaar to access rations, MGNREGA, social pensions, SIM cards or bank accounts for the first time. And 61 per cent beneficiaries trust that Aadhaar prevents 'others' from accessing their benefits.

These numbers would become even better if some of the problems that people face get reduced, which must be the priority of the government and other stakeholders too. We can then truly say that Aadhaar is not for the happiness of many, or the majority, but for everyone.

Chapter 7

PARANOIA ABOUT DATA SECURITY

*Among the Internet's many gains for humanity, decreasing paranoia
has not been one of them. Anything from that lump under your
armpit to what's lurking in the sea—just type it into a search
engine and watch your nerves explode.*

—John Niven, Scottish author

Aadhaar touches the most fundamental of human values: our
need for identity, the need to guard our privacy, the need to
preserve self-interest and agency, and the need to look out for the
less privileged in society, among others. Naturally, that makes us
anxious about this new-fangled ID that is an online artefact—not a
card, certificate or some other token that we understand as an ID.
Further, the pundits have told us to be distrusting of it for a variety
of reasons that seemed plausible at some early point. However, once
the mind is made up, human beings often resist change, even if it
is to be on the basis of a Supreme Court judgement, a scientist's
testimony or a cogent argument. In the case of Aadhaar, they had
the option to join any of the four opposition camps.

People in the first camp are worried about possible loss of
data entrusted to UIDAI. In the second camp, people worry about
profiling of individuals made possible due to Aadhaar linking. It
isn't linking, as we shall see, but seeding of the Aadhaar number
in some databases. Worrywarts want this practice to stop.

The third camp consists of people who fear that the government,
with the aid of UIDAI, has acquired the capability to watch citizens
because of Aadhaar's authentication logs. Please note that this camp,

afraid of surveillance, is different from the second, which consists of people afraid of profiling. These distinctions are necessary if our thinking isn't to get befuddled by vague fears and unanswered questions. The last camp hosts people who believe that just the disclosure of the Aadhaar number increases their vulnerability in the digital world.

It is necessary to understand the ideological beliefs and arguments of all the four camps to be able to separate fact from fiction. Let's hear them one at a time, quietening the other three voices till it's their turn.

The Breach Brigade

People in this first camp are rightly worried. Data breaches happen and the more valuable the data, the more likely it is that someone would attempt to steal it. You protect your data like you protect other things of value. You lock your home when you leave. Valuables such as cash or jewellery, you put in a locker—and take the key with you. For the truly expensive stuff, however, you rent a locker at the bank.

UIDAI built the equivalent of a bank's locker for you. It's protected by physical security, has stringent access protocols to prevent intrusion and it keeps the data under locks that are virtually impossible to pick. Your biometric data is safe in UIDAI's vault. It's never allowed to be let out. Your Aadhaar number is merely a kind of receipt. In fact, you can even put an additional lock by instructing UIDAI not to allow any authentication attempts on your Aadhaar, as discussed earlier. This is called locking of your Aadhaar number.

However, that doesn't prevent the distress created by stories of Aadhaar data breaches. The captions in the press are scary and designed to create the impression that something serious has happened, and personal and sensitive information of citizens was hacked.

For example, there was the data leak in the Employees Provident Fund Organisation (EPFO), which was presented as an Aadhaar data leak because the data records of EPFO beneficiaries contained their

Aadhaar numbers too.[147] Reacting to such distorted presentation of facts, a wit asked an interesting question: if cash gets stolen, can it be termed as an RBI data leak because the currency notes carry RBI serial numbers?

Like WhatsApp forwards, such stories prove to be fake time after time. It isn't easy to spot fake stories. Even experts are prone to mistaking the fake ones for the real. For instance, a digital security company, Gemalto's CEO, Philippe Vallée personally offered his 'sincerest apologies on the grave error' after erroneously publishing a report announcing a huge Aadhaar data breach![148]

Should Aadhaar Numbers be Published? When the next story of Aadhaar data breach happens, here is what you need to keep in mind. The RTI Act mandates that government agencies should proactively publish the list of beneficiaries of their welfare programmes on their websites. Section 4 specifies the details and manner of publication of this information.[149] Accordingly, the lists of MGNREGA beneficiaries along with all the details such as the panchayat, the project, the wages paid and their bank accounts were being routinely published on the websites of state governments. Nobody complained about it. Such disclosure was considered good for transparency and in accordance with law.

---

[147]Scroll Staff, 'Employees Provident Fund Organisation Shuts Down Aadhaar Seeding Website after Data Leak,' Scroll.in, 2 May 2018. Available at: https://scroll. in/latest/877682/employees-provident-fund-organisation-shuts-down-aadhaar-seeding-website-after-data-leak. Last accessed on 7 May 2020.

[148]PTI, 'Aadhaar Data Breach Report: Digital security firm Gemalto issues public apology,' *The Hindu BusinessLine,* 27 October 2018. Available at: https://www.thehindubusinessline. com/info-tech/aadhaar-data-breach-report-digital-security-firm-gemalto-issues-public-apology/article25347670.ece. Last accessed on 7 May 2020.

[149]Subsection 4(1)(b)(xii) and (xiii) of the RTI Act:
Every Authority shall publish...

> *(xii) the manner of execution of subsidy programmes, including the amounts allocated and the details of beneficiaries of such programmes;*
>
> *(xiii) particulars of recipients of concessions, permits or authorizations granted by it;*
> Section 4(2): *It shall be a constant endeavour of every public authority to take steps in accordance with the requirements of clause (b) of subsection (1) to provide as much information suo motu to the public at regular intervals through various means of communications, including internet, so that the public have minimum resort to the use of this Act to obtain information*

As the Aadhaar case was being argued in the Supreme Court, this practice of publication of Aadhaar data was 'discovered' by some activists and exposed as a massive Aadhaar data leak. It was implied that the Aadhaar system was, therefore, completely unsafe!

However, on closer examination, we observe that publication of details of beneficiaries cannot be flogged as a data breach, however much the breach brigade would like everybody to believe so. Firstly, Aadhaar is not a secret or confidential number. It is a random number bereft of any intelligence that does not contain any sensitive personal data. It is attached to an individual much like her name. This linkage is unique because one Aadhaar number is not linked to two individuals, nor does any individual have claim to two such numbers.

However, Section 29(4) of the Aadhaar Act prohibits the publication of Aadhaar numbers except for the purposes specified in Regulations. The Regulations, in turn, provide that no entity shall make public any database or record containing Aadhaar numbers of individuals, unless these numbers have been 'redacted or blacked out through appropriate means, both in print and electronic form.' Thus, if the authorities do publish the information, the Aadhaar number should be either partially or fully masked in that publication.

The purpose of these restrictions is that while the Aadhaar number itself is not confidential, publication of Aadhaar numbers 'in public records' will make it easy to collate information about a person. Collation of data, unfortunately, has become relatively easy in the digital world even otherwise.

On the other hand, the objective behind RTI is to ensure transparency in the functioning of public authorities and to enable social audit of their programmes. Therefore, public authorities are under obligation to provide the information on record with them, unless it is expressly prohibited under the Act. If a person seeks the details of beneficiaries of any scheme under the RTI Act, public authorities are duty-bound to provide the information they have, including the Aadhaar numbers. Exemptions under Section 8 of the RTI Act will not help the authority to hold back the information.

What constitutes the details for the purpose of publishing on the websites? Should Aadhaar number be included in the details of the person or not? This is essentially a question of balancing transparency of public records under the RTI Act, for instance, and the privacy of individuals enunciated in Section 29 of the Aadhaar Act.

My personal view is that the last four digits of the Aadhaar number can be published and the first eight digits can be redacted/masked. This will satisfy the provisions of both the RTI and the Aadhaar acts. However, to say that disclosure of Aadhaar numbers by authorities in compliance with the law of the land constitutes a data breach or data leak is preposterous. If we don't want some data in public domain, we must amend the law that requires its disclosure, rather than fixing Aadhaar.[150]

It is amazing that some of the same activists who fought for RTI and transparency have now found reason to howl that beneficiary details are being transparently displayed on the websites of the departments and their agencies. It should be clear that such 'data leaks' have nothing to do with the security architecture of the Aadhaar system which has a very high level of encryption. It has been categorically asserted by the government that there hasn't been a single instance of breach or leak of data from UIDAI.

Consider this: the entire list of electors in India with their address and other personally identifiable information is available online. Under the NPR, it is mandatory to publish the list of residents publicly. Telephone directories have such information. None of these constitute any data breach and we are also comfortable with privacy implications of these systems. But with Aadhaar, the perception is different.

The story that a manager at some Common Service Centre (CSC), handling Aadhaar enrolment, had sold his credentials for

---

[150]R.S. Sharma, 'There Has Been No Aadhaar "Data Leak". Chances of That Happening Are Almost Zero', *The Economic Times*, 9 March 2017. Available at: https://telecom.economictimes.indiatimes.com/tele-talk/there-has-been-no-aadhaar-data-leak-chances-of-that-happening-are-almost-zero/2341. Last accessed on 7 May 2020.

about ₹500[151] created widespread panic. It was publicized as 'total data leak' of a billion people. But a little understanding of how the system works would assure anyone that it was nothing of the sort.

Residents have the facility to download a digitally signed copy of their Aadhaar letter and print it whenever they want. To do so, the resident must provide her Aadhaar number, full name and PIN code, and then enter a security code (CAPTCHA) to request an OTP on her registered mobile number. If you forget the Aadhaar number, you have alternatives that require you to provide the date of birth, enrolment number, etc. If you don't have access to an internet-connected computer, you may visit a CSC or one of UIDAI's contact centres, where an authorized operator can help you do the same thing.

What happens if anyone illegitimately acquires the credentials of a CSC operator? They can help a resident print the Aadhaar letter! But it is the resident who must tell the Aadhaar number to the operator or his proxy and provide further details such as mobile number or PIN code as required. That people need to download their Aadhaar letter is obvious and clearly this facility has been widely used and hasn't created any issues. Roughly four lakh such downloads have happened every day and cumulatively their number is more than a billion.

The probability of guessing someone's Aadhaar number, given the fact that he or she has one, is less than one in 50 billion because less than 2 per cent of a billion possible numbers are actually allotted.

But what if you do obtain somebody's Aadhaar number and other details that allow you to print her card too? Well, it gets you a printout, nothing more. The Aadhaar letter is designed to be useless by itself. 'To establish identity, authenticate online,' is prominently printed on every such letter. The Aadhaar number can be used for identification only in conjunction with online authentication, using the resident's fingerprint or iris scan. Or in a weaker form of authentication, together with the OTP sent by

---

[151]Rachna Khaira, 'Rs 500, 10 Minutes, And You Have Access to Billion Aadhaar Details,' *The Tribune*, 3 January 2018. Available at: https://www.tribuneindia.com/news/archive/nation/rs-500-10-minutes-and-you-have-access-to-billion-aadhaar-details-523361. Last accessed on 7 May 2020

UIDAI to the resident's registered mobile number for each attempt at authentication.

So, if a person shares her credentials for printing Aadhaar letters with another for a consideration, as happened in our story, can it be called a breach that exposes the weakness of the system? It would be like someone sharing their Wi-Fi password being termed as successful hacking of the internet! This episode was part of a sustained campaign launched to discredit Aadhaar.[152]

Is Data Linking Dangerous? There are two broad categories of linking. The first is linking of your mobile number with your Aadhaar and the second is seeding of other databases that have information about residents with their Aadhaar number.

Having the mobile number on record enables a communication channel between UIDAI and the resident, which is used to provide an alert whenever the Aadhaar number is used for authentication, successful or otherwise, such as in a banking transaction or while submitting income tax (IT) returns online. Even if your Chartered Accountant submits the IT return, the OTP is sent to you and, therefore, the return is only filed with your knowledge and on your behalf.

Thus, the mobile becomes your de facto digital identity and provides another convenient way to authenticate you for availing various e-services. This mode of authentication has become important for financial inclusion, such as through Unified Payments Interface (UPI) or Aadhaar Pay.

Data linking by seeding Aadhaar in other databases is intended to prevent leakages in benefit delivery programmes involving entitlements and subsidies and to ensure compliance with the law of the land. Aadhaar enables this by cleaning up databases of duplicates and ghosts.

For instance, the Income Tax department needs to ensure that everyone assessed to tax is uniquely identifiable. Indeed, this was

[152]R.S. Sharma, 'There Has Been No Aadhaar Data-Breach Till Date,' *The Economic Times*, 15 January 2018. Available at: https://economictimes.indiatimes.com/news/economy/policy/there-has-been-no-aadhaar-data-breach-till-date-rs-sharma/articleshow/62500216.cms. Last accessed on 7 May 2020.

sought to be achieved by issuing a PAN card to each assessee. Unfortunately, some unscrupulous elements managed to get more than one for themselves. Even if the department had succeeded in its objective of one card per assessee, another card in the name of a non-taxpayer could always be obtained and used for tax evasion or creation of benami accounts and transactions.

Clearly this problem required a technological solution that covered the country's full resident population. In other words, the Income Tax department needed to do its own UID project for the country to achieve its end objectives. If PDS or MGNREGA or LPG subsidy and other such programmes similarly attempted an Aadhaar-scale infrastructure, what a colossal waste of public money it would be!

The alternative solution is simply to seed the Aadhaar number in the same table that has the resident's other personal identification, such as name and PAN number or ration card number or MGNREGA number. All accounts that have no Aadhaar number can then be removed following due diligence and opportunity to the individual to remove the deficiency. This applies to individuals only and not to corporate entities, which may have a PAN number but not an Aadhaar number.

The presence of Aadhaar in one field of a database thus ensures elimination of duplicates and easy authentication of the individual through fingerprint, iris scan or mobile OTP. In other words, it is a solution at virtually no cost to the government or business entities that need to identify its users because Aadhaar is created as a public good.[153]

Aadhaar is like a detergent that cleanses, but not everybody likes a good wash. Now that it is cleaning other systems to uncover benami properties and tax evasions, it seems to have become too dangerous for some people and is being demonized as a great threat to freedom and privacy.

---

[153]In economics, 'Public Good' has two connotations, both of which apply to Aadhaar; it is a service provided without profit and it is a good that is both non-excludable and non-rivalrous, in that individuals cannot be effectively excluded from use and where use by one individual does not reduce availability to others.

Concerns Relating to Profiling and Data Aggregation

The privacy vulnerabilities created by online social networks have been in the news lately, having us justifiably worried. These vulnerabilities have been studied in terms of graph theory in recent literature.[154] This branch of mathematics is used to model pairwise relations between objects and extensively used in the development and deployment of social media platforms.

However, there are some crucial differences between this discussion about online social networks and what we know about Aadhaar. Social networks may be modelled as a directed graph, where there is a direction associated with the relationship between objects, or undirected graphs. Aadhaar, on the other hand, is not a graph because it maintains no linkages.

Consider the difference between Facebook and email. Facebook maintains a relationship between individuals, which is known to Facebook itself and could be visible to third parties or other users of the system. The same holds true for other social networks such as Twitter or Instagram. By crawling from one node to another along the linkages, it is possible to completely map out the relationships between the users. Friend suggestions by Facebook is derived from such linkages. Further, by reading the messages, it is also possible to profile them more comprehensively.

There is no such relationship maintained among email systems. When an email message is sent from one user to another, it traverses a path that leaves identification marks and timestamps in the message header, but that's all. To capture and analyse a limited set of such messages, you must hack into the email systems or the network itself. Even then, the federated architecture of email doesn't lend itself to comprehensive profiling of all users.

Aadhaar has a federated structure like email, but with many more safeguards. For instance, one can often guess some information from an email address such as chairman@trai.gov.in, but precious

---

[154]L.A. Cutillo, R. Molva and M. Onen, 'Analysis of Privacy in Online Social Networks from the Graph Theory Perspective,' 2011 IEEE Global Telecommunications Conference—GLOBECOM 2011, Houston, TX, USA, 2011, pp. 1-5.

little from 8ku9f96xwcfc2cst@protonmail.com because the latter has no structure. Aadhaar is like the latter: a number formed with 12 random digits. Most Aadhaar number lookalikes that you could construct from 12 digits are either invalid or unassigned.

Email carries text, images and other graphics. Aadhaar authentication carries only the user's Aadhaar number and basic information (already with UIDAI) from which nothing can be guessed about its purpose. Such a request could come from a bank or a government office or a telco or anybody, and without revealing who is asking the question. Unlike online social networks, where the communication and/or linkage is often public and visible, in the case of Aadhaar, there is no linkage and the communication is encrypted between UIDAI and Authentication User Agency (AUA). UIDAI connects only to the AUAs, which are a few hundred entities that provide UIDAI's authentication services to others. Further, the enrolment data at UIDAI is stored in a well-protected data centre that has been hardened against intrusions and encrypted.

Aadhaar, thus, has none of the risks associated with social networks that ignited fierce privacy debates in recent times. Yet, in the minds of the people, the two are conflated. They hear Cambridge Analytica and think Aadhaar.

Do Authentication Footprints Constitute Surveillance?

In 1949, George Orwell published the novel *Nineteen Eighty-Four* (often published as *1984*) referring to the year—then in the future—when people get entrapped in a world of perpetual war, surveillance and propaganda. It's a thought-provoking and deeply disturbing view of the world that we could invite upon ourselves. But Orwell's novel is about as relevant to understanding Aadhaar as his equally brilliant parable *Animal Farm* is insightful about the lives of pigs.

In the world we live in today, we leave footprints from all digital transactions. Google knows about the searches we conduct and places we visit; Facebook knows about our lives because we ourselves keep it updated and Amazon knows what we buy, what we are considering and what we might buy if it's positioned in our view.

The airline knows when we check in or board a flight. The telephone company has the logs of all our calls, including the numbers called, the duration of our conversations and the location we called from. The law allows some of this information to be shared with law enforcement—with safeguards, of course—because it helps control crime or with emergency responders when we call for help.

In a way, we are constantly under surveillance. We have allowed the government to do this under certain circumstances. But we have allowed almost unfettered surveillance by corporates that exceeds the reach and power of governments today. Experts warn us of the dangers of online surveillance. Yet, we ignore those warnings the same way that we ignore the doctors who tell us not to consume too much sugar or spend that much time on the couch. That doesn't, however, make us any complacent about the imaginary dangers of vaccination, the cancers caused by cell tower radiations and the ill-effects of birth control pills.

So, does Aadhaar authentication leave digital footprints that can be used by UIDAI or others to profile a citizen's life? No. Because UIDAI does not know the purpose or the location of an authentication request. It has no idea about the internet protocol (IP) address of the machine from where the request is made.

Think of authentication request to Aadhaar in the same way as you would consider a visit to your Facebook page using an anonymous proxy server from a clean computer just one time—and then closing the browser window without any activity.

The Supreme Court of India that unequivocally read into the Constitution a citizen's right to privacy, heard privacy arguments against Aadhaar and determined that both Aadhaar and its authentication services are secure and do not violate a citizen's fundamental right to privacy.[155] Furthermore, it concluded that Aadhaar does not enable mass surveillance because information required for a surveillance system is not collected during authentication: the logs contain just enough information to prove

---

[155]Supreme Court of India Writ Petition (Civil) No. 494 of 2012, Pg 153.

an instance of authentication, but not who made that request or for what purpose.

On the other hand, authentication empowers you. Nobody else can take away the entitlements of the poor through proxy. If I am entitled to subsidized ration, then only I can withdraw my entitlement from the ration shop. Even the shopkeeper cannot now deny my entitlements, as it is only I who can authenticate myself.

Does Disclosure Increase Vulnerability?

That brings us to the fourth camp, which believes that the disclosure of Aadhaar increases a citizen's vulnerability in the digital world. This is a widely misunderstood aspect, sadly even among the 'expert' supporters of Aadhaar! It should be clear by now that disclosure of Aadhaar, like the disclosure of your bank account printed on your cheques or the telephone number or email address displayed on your visiting card or even on the internet, is not a risk.

So, what exactly does mere knowledge of an Aadhaar number allow? Nothing, to be precise. Practically, my Aadhaar number cannot be used to access my bank account or any other digital resource, which is password protected. Aadhaar number will normally not be the username in any system. However, even if I was foolhardy and used my Aadhaar number as my bank's user ID, it cannot be a password for any system. Hence, knowledge of my Aadhaar number does not make access to any of my digital resources any easier. Can you authenticate me? Firstly, it is not possible. Even if you did, you will be attempting to commit a crime punishable under the Indian Penal Code (IPC) as also under the Aadhaar Act.

In one of the interviews, given to the online magazine, The Print,[156] I had reiterated the assertion that no harm can be caused to me if I disclosed my Aadhaar number. When one Twitter user challenged me to walk the talk and publish my Aadhaar details

---

[156]Regina Mihindukulasuriya, 'What Harm Can You Do to Me If You Have My Aadhaar Details, Asks TRAI Chairman R.S. Sharma,' The Print, 27 July 2018. Available at: https://theprint.in/governance/what-harm-can-you-do-to-me-if-you-have-my-aadhaar-details-asks-trai-chairman-r-s-sharma/88798/. Last accessed on 7 May 2020.

if I had so much trust in the system, I decided I should have the courage to act on my belief. I disclosed my Aadhaar number on Twitter. While I am an impulsive person at times, this tweet was not an impulsive one.[157] I hadn't expected it, but the tweet became viral. However, nobody could demonstrate any harm, either then or in the months that have followed since.

People did make fun of me and issued threats that alarmed my well-wishers, but all of it was ultimately of no consequence. I was trolled for more than a week. I maintained a brave face, but those days were trying for me. While I did not want to see the Twitter messages from the trollers, I was drawn to them. Some opined that I was going to have an ignominious exit after four decades of a career in public service.

I received emails advising me to recant the assertion and salvage my reputation, such as was left of it. But how does one take back a number? Even if I deleted the tweet, the number was reproduced elsewhere in the public domain. There was no going back.

Of course, when the hackers tried to show that they could cause me some harm, I did take a few precautionary steps. For example, I changed my passwords and hardened them on my bank accounts, social media accounts and email accounts. Yet, when hackers are out 24×7 to cause harm, one feels vulnerable. One does not know the source of the attack.

The real danger was that even if the hacked information had no connection with Aadhaar, it would be shown as hacking enabled by my Aadhaar disclosure. Just as children would sometimes blame a pet for their mischief. In this case also, my Display Picture (DP) on WhatsApp was claimed as having been extracted due to my Aadhaar disclosure. Even a child knows that you can see the DP if you know the mobile number. Unfortunately, my mobile number is not a state secret. In any case, the so-called disclosure had nothing to do with my Aadhaar. After about 10 days of this very public spat,

---

[157]R.S. Sharma, 'Why I Gave Out My Aadhaar Number', *The Indian Express*, 1 August 2018. Available at: https://indianexpress.com/article/opinion/columns/rs-sharma-aadhaar-number-challenge-trai-chairman-twitter-5283781/. Last accessed on 7 May 2020.

I wrote about the experience in *The Indian Express*.[158]

This experience brought some important lessons for me. First, it confirmed my belief that disclosure of Aadhaar does not increase digital vulnerabilities, which was tested and confirmed. This should be of comfort to those uneasy about disclosing their Aadhaar because of the campaign of disinformation that has been spread.

Another lesson emerged out of the need for a theoretical validation of my belief. When I discussed this issue with a few mathematicians and others good at graph theory, I found that Aadhaar does not increase either the privacy or security vulnerabilities already present in the digital world.

One such person was Professor Manindra Agrawal of IIT Kanpur, a mathematician of shattering eminence. He investigated the three main risks usually presented against Aadhaar—surveillance, forgery and database attacks—and found that Aadhaar does not increase either the privacy or security vulnerabilities already present in the digital world. His testimony was also submitted to the Supreme Court in the PIL. The third lesson I learnt is that social media, especially Twitter, is not an appropriate platform to discuss public policy issues.

Nevertheless, to address the concerns of those who are afraid of number disclosure, UIDAI has come up with the concept of virtual ID, which ensures that even when you authenticate, you need not give your real Aadhaar number, but a 16-digit number provided to you by UIDAI. Now, no entity can do a reverse mapping to your Aadhaar number from the virtual ID. You may even change the virtual ID when you want to.[159]

It isn't only the national institutions that have scrutinized Aadhaar. The experiment has been watched by other countries and international institution, including the International Telecommunication Union (ITU), which created a guide for Digital

---

[158]R.S. Sharma, 'Truth Fears No Trolls,' *The Indian Express*, 9 August 2018. Available at: https://indianexpress.com/article/opinion/columns/rs-sharma-aadhaar-number-challenge-twitter-trai-chairman-data-protection-5298308/. Last accessed on 7 May 2020
[159]Generate/Retrieve Virtual ID. Available at: https://resident.uidai.gov.in/vid-generation. Last accessed on 11 May 2020.

Identity Roadmap[160] taking note of India's Aadhaar system. Some other countries are pursuing similar programmes of their own.

The architectural choices for Aadhaar were carefully made to achieve the programme objectives. They were also made to protect the privacy and security needs of the users. That is how the Aadhaar protocol could become the core around which much else can be built and is indeed being built, expanding the benefits to society without any reduction in security and privacy.

However, Aadhaar should not be considered as the end of innovation in identity systems. Technology continues to evolve and bring newer threats to society. We must remain ever vigilant and support further developments to mitigate against those threats.

---

[160]Digital Identity Roadmap Guide. Available at: https://www.itu.int/pub/D-STR-DIGITAL.01-2018#:~:text=The%20Digital%20Identity%20Roadmap%20Guide,a%20 National%20Digital%20Identity%20Framework. Last accessed on 15 June 2020.

Chapter 8

THE STRUGGLE WITHIN

It is natural that we face obstacles in pursuit of our goals. But if we remain passive, making no effort to solve the problems we meet, conflicts will arise and hindrances will grow. Transforming these obstacles into opportunities is a challenge to our human ingenuity.

—His Holiness, The Dalai Lama

When I joined UIDAI in July 2009, never did I imagine that a government project would face so many challenges and existential crises, that too from within the government itself.

Various high-level stakeholders took great exception to the Aadhaar project from the word go. We seemed to find ourselves at loggerheads with a variety of offices—be it the Home Ministry, the Planning Commission or the Finance Ministry. This is not to mention the number of judicial challenges and loud protestations from NGOs.

Shortly after the project was initiated, Dr Manmohan Singh instituted the PM's Council on UIDAI. The council convened for the first time on 13 August 2009 and approved UIDAI's Strategy Document. Going further, the Cabinet Committee on UIDAI (CCUIDAI) was officially constituted to take decisions about the UID project. CCUIDAI would later become the forum for opposition to the project.

Somebody has interestingly summarized four arguments against any project or idea in the government. Those of us in the government would have come across all of these, at various points in time, in our career. They are formidable arguments and often one has no answer, as these are just arguments and the person advancing it does not have to produce a rigorous proof in support thereof.

It has already been done: This is the first idea and was forcefully applied in our case. When we talked about creating unique IDs for the residents of India, people would come back and say: 'So, what is new about it? It has already been done. We have various IDs in our country: ration card, BPL card, Voter ID card, passport and MGNREGA job card. When everybody already has one or the other ID, what is the need for another one?'

The second argument goes as follows. Even if you do create unique IDs, they would serve no great purpose and the expenditure would be a waste. The problem of duplicates and ghosts, which you are claiming to resolve by creating such IDs, are not serious problems for which you should undertake the massive exercise you propose. Hence, the assertion goes: *it is not worth doing.*

The third contention is: you cannot do it. This was frequently used in our case. As no country in the world has ever done it (creating UIDs at such a massive scale), therefore you cannot do it. This was thrown at us by many in the government besides other outside experts.

The last and the most potent argument was: you have no authority to do it. Creation of IDs is the legitimate function of the Home Ministry and UIDAI has no authority over it. This was argued even though the same government had created UIDAI after three years of deliberations by EGoM, with the specific mandate of creating unique identity for the residents of India.

The Resistance

While the project had to bear different kinds of opposition from outside the government, we are going to describe here some of the challenges we faced from within the system.

The Turf Challenge: The idea that the MHA is the designated agency to undertake this ID project, was the root of this turf battle. Around the same time when the UID project was started, the RGI was in the process of creating the NPR. The NPR, as mentioned earlier in the book, was a diluted version of NRC, envisaged under the Citizenship Act, 1955. Along with the preparation of the 2011 Census of India, the RGI had created the family schedules, also known as NPR schedules. These schedules had collected the details of family members of the people surveyed for the 2011 Census.

Beginning the work on our project, we were given the option to take all the NPR schedules in paper-form and proceed from that point onwards. The MHA/RGI, who was implementing NPR, did not have any intention to take the biometrics of the people. We did not agree to take these schedules, as our model was to enrol people on a voluntary basis and we had no intention of digitizing all the NPR schedules. The conflict started from there.

Somehow, probably as a reaction, the MHA/RGI decided that they would also take biometrics and made changes in their rules, etc. The criticism that two agencies of the same government were collecting biometrics leading to duplication of efforts and wastage of public money became a controversy and a point of conflict within the Government.

It was argued that government agencies at the grassroots were collecting resident data for NPR enrolment. On the other hand, UIDAI too was collecting data through its multiple registrars in a so-called 'haphazard' manner without any system of verification—or so the accusation went.

The opposition was twofold. First, public funds were being spent twice to collect the same information. Second, and more centrally, it was believed that the NPR data was being collected in a robust manner, eliminating the possibility of any fraud in enrolments, whereas UIDAI was enrolling residents through private agencies, with 'no accountability', thus compromising the credibility of the data itself.

We, at UIDAI, would not have objected to utilize the data collected for NPR, as there were similarities in the data being procured by each, but we found NPR's multistage process slow and inefficient. The problem of duplication of the work was resolved by CCUIDAI by dividing the states into NPR and UIDAI states and devising a protocol as to how both agencies could work together without causing major duplication of work. The above question of credibility and UIDAI's trepidation about NPR's methods stood legitimized when NPR's enrolments fell far behind that of UIDAI's and we continued to be allocated more and more states and Union Territories.

UIDAI vs MHA, Again: One of Aadhaar's first near-death experiences came even before the project was launched. The PM's Council permitted us to conduct a pilot of 100 million IDs, which we expeditiously completed in about 10 months. Having effectively proved our model, we needed to quickly expand the scope of our project to cover more population. Given our eagerness to ride the wave of our momentum and spread of enrolment, we sought approval to enrol another 100 million residents, from the Ministry of Finance (MoF), as it was more quickly accessible than the PM's Council. The MoF duly approved the proposal. Beneficial as this was, it didn't sit well with the MHA that took exception to the fact that this approval was granted by the MoF without the express consent of CCUIDAI.

This wasn't the last time we had issues about NPR, with more problems coming with the change in administration in 2014. The new PM constituted a group of ministers to study the issue of harmonizing Aadhaar and NPR and it was tentatively decided that the RGI would be the sole source of data for UIDAI. In other words, UIDAI would only be a back-end data-processing agency that would issue Aadhaar numbers.

There were many ups and downs in this continued struggle and at times, it seemed that we would have to close up shop. However, what worked in our favour was our speed of execution. We had promised, in our first Strategy Document, to complete 600 million UIDs by the end of 2014.[161] We delivered these numbers by March 2014, a full nine months before the due date! To be fair, we delivered these numbers not because we were more competent

---

than others, but because our operating model made the enrolment process scalable. We also did not have sequential and multistage processes involving multiple parties and huge coordination costs that can slow down the progress of any project. We had coined a word 'AUM'—where A stands for asynchronous, U for unbundled involving minimal dependencies on other systems and processes and M for minimalistic. We have extensively used these basic approaches wherever feasible, in the UID project.

Planning Commission Demands Accountability: Another blow to Aadhaar came at the hands of the Planning Commission that raised concerns regarding the accountability structure of UIDAI. Ganga was working as our financial advisor. As the team was still relatively small, she had other vital responsibilities in UIDAI. The Planning Commission insisted that the purview of her work should not extend beyond that of a financial advisor. Consequently, Ganga was divested of all other duties. Unfortunate as this was, we recognized the merit in picking our battles.

Existential Threat: In March 2013, I left UIDAI to join the government of Jharkhand as Chief Secretary. I was eager to leverage Aadhaar in Jharkhand for the many projects that I had initiated. During this tenure, I unleashed Aadhaar's fullest potential for several kinds of service delivery.

First, I initiated the Aadhaar Enabled Biometric Attendance System (AEBAS) in the Ranchi Secretariat, which was eventually extended to the districts. We then took on the monumental task of seeding Aadhaar in various databases with a view to introduce DBT in programmes involving payments or subsidies such as the NSAP, MGNREGA and scholarships. We even employed Aadhaar's biometric authentication capabilities in the distribution of foodgrains through the PDS and authentication of parties for property registration. Clearly, Aadhaar was working well in the state leading to significant acceleration of its coverage. Witnessing Jharkhand's journey with Aadhaar, other states similarly began to ramp up their enrolment efforts.

I had originally planned to continue in Jharkhand till my

superannuation on 30 September 2015 and settle down in Ranchi, a town I love. But as it so often happens, things did not work out as planned. After spending about a year in Jharkhand, I was longing to go back to the GoI. With some effort, I joined the Centre as Secretary in Delhi in the Department of Electronics and Information Technology (DeitY, present-day Ministry of Electronics and Information Technology or MeitY), a part of the Ministry of Communication and Information Technology, on 1 May 2014.

When I returned to Delhi, however, the scenario had changed completely. Within days of my joining the department, the landmark change of government took place. This was to be the direst existential threat to the Aadhaar project since its inception.

The Bhartiya Janata Party (BJP) in its manifesto, had promised to review the Aadhaar programme with implicit intent to shut it down altogether during the whirlwind election campaign. Shri Narendra Modi had launched the project in his state way back in 2012 when he was the CM of Gujarat. I know this because on 1 May 2012, we had enrolled him for Aadhaar during the Gujarat Day function in Gandhinagar.[162] But given the party's rhetoric on the subject in the run-up to the elections, I was extremely fearful of the project being abandoned. These fears were further stoked when, just days after taking office, the government saw it fit to dissolve CCUIDAI on 10 June.

As there was no official government order on the future of Aadhaar, I had called a meeting of all the agencies using Aadhaar in the performance of their duties. This was in my capacity as Secretary of DeitY, the department that was essentially in charge of e-governance in general, and obviously Aadhaar applications fell within the purview of e-governance. This meeting was held on 18 June, the day after I had issued a statement appealing for completion of Aadhaar's seeding process, which appeared in *The Indian Express* as follows:

---

[162]Express News Service & PTI, 'UID Project Launched in Gujarat', *The Indian Express*, 2 May 2012. Available at: https://indianexpress.com/article/cities/ahmedabad/uid-project-launched-in-gujarat/. Last accessed on 15 June 2020.

DeitY Secretary R.S. Sharma told *The Indian Express* that it was for the political executive to provide a direction on how Aadhaar and its twin project, the National Population Register, will move. 'We have called a meeting of those states where the extent of rollout of these cards has almost been completed, to decide their future course of action,' Sharma said.[163]

Ravi Shankar Prasad, our new minister, soon called me on phone, livid. He asked why I had called such a meeting when the government was planning on abandoning it, thus inadvertently justifying my fears. I patiently responded by saying that as long as there was no official order to that effect, Aadhaar continued to be a government programme and it was my duty as DeitY Secretary to review all programmes relating to e-governance. Seeing the minister's unhappiness, we decided not to issue a press note on the meeting that day.

Meanwhile, the PM was conducting review meetings with various government departments, clubbing certain departments, along with their respective secretaries and ministers, into single meetings. DeitY's meeting happened to be clubbed with the departments of Posts, Telecom and Information and Broadcasting (I&B), and on 23 June I presented my case. Considering our minister's position, I did not specifically use the term 'Aadhaar' in my presentation (which, I believe, was the reason it was cleared by the minister). However, the focus of my presentation was the proposal for the country to develop various platforms such as identity and payments to provide horizontal services. I had secretly planned to talk about Aadhaar specifically, if given even half a chance.

The presentation started around six o'clock in the evening and continued for more than two hours at the illustrious 7, Race Course Road (presently renamed Lok Kalyan Marg) address. When we came to the subject of creating platforms, I had wanted to subtly broach

---

[163]Subhomoy Bhattacharjee, 'Aadhaar Future at Stake, Govt Seeks Meeting with States,' *The Financial Express*, 17 June 2014. Available at: https://indianexpress.com/article/india/india-others/aadhaar-future-at-stake-govt-seeks-meeting-with-states/. Last accessed on 11 May 2020.

the issue of Aadhaar. However, in a flash, something came over me, and I began to wax lyrical about Aadhaar.

I told the PM that if there was one programme that was transformational from the viewpoint of governance, this was it. I told him that while the government would spend only about ₹8,000–9,000 crore (₹80–90 billion) on Aadhaar, it would ultimately save around ₹50,000 crore (₹500 billion) annually, just by way of eliminating duplicates and fakes from the various beneficiary databases. When asked, I enumerated the various individual/family beneficiary-oriented schemes (PDS, MGNREGA, NSAP, scholarships, LPG subsidies, etc.), where the scope for eliminating duplicates and fakes was immense. I pointed out that the country—both Central and state government—was spending around ₹5 lakh crore (₹5 trillion) annually on these schemes, while the average number of duplicates and fakes were at least 10 per cent (as a conservative estimate), thus, eliminating these would result in phenomenal savings.

At the end of that presentation and the long conversation on Aadhaar that followed it, the PM, to my surprise, asked if I had implemented some kind of attendance system for the government employees in Jharkhand. When I replied in the affirmative, he asked me whether I could implement the same in the central government as well. A second affirmative answer led to deadlines being set as early as October of the same year (2014), for the attendance project's commencement. But I made it perfectly clear to him, at the meeting itself, that my system was fundamentally reliant on Aadhaar authentication.

I was on top of the world after this meeting. I wasted no time in calling Nandan to share with him the discussion that had just taken place regarding Aadhaar. We saw a ray of hope for Aadhaar's survival. Logically, the PM asking me to introduce a biometric attendance system based on Aadhaar implied that Aadhaar was going to stay. However, I was not sure whether the PM had considered that there are several widely used biometric attendance systems that don't require Aadhaar specifically. Thus, our hope remained but a ray.

Thereafter, I met with Nandan for lunch at his Safdarjung Lane residence in New Delhi, where we discussed a variety of issues. The probability of Aadhaar surviving, unsurprisingly dominated the conversation. I quickly suggested to Nandan that he meet with the PM to cement Aadhaar's future in India. My view was that, while there was significant discussion around Aadhaar in the meeting of 23 June, it still required a further push, and only Nandan had the stature to really convince the PM into continuing the project. Nandan, in his humility, was reluctant to do so, but when his wife Rohini supported me, he was press ganged into agreeing.

Nandan eventually met with the PM on 5 July 2014, to discuss the fate of Aadhaar. Though I do not know the comprehensive details of that discussion, Nandan told me after the meeting that the PM was 'positive' on the idea. And, when the PM asked him to suggest someone to drive the project, he suggested my name, being the ex-DG and the current Secretary, DeitY.

Since then, Aadhaar has had a new lease of life. And what a life it has been! There are currently more than 1.25 billion people with Aadhaar numbers and its application has stepped into innumerable and unimaginable areas, all thanks to our PM, Modi.

Surviving NAC: Another strong opposition to Aadhaar came from the National Advisory Council (NAC). The first NAC was constituted in 2004 by then PM, Dr Manmohan Singh, to implement the National Common Minimum Programme (NCMP). The NAC was reconstituted in 2010 under the chairpersonship of Sonia Gandhi. Other NAC members included Mihir Shah, Dr N.C. Saxena, Aruna Roy and Jean Drèze. They were in the process of drafting the Food Security Bill.

We had been pitching for using Aadhaar as a means of ensuring accountability in the food distribution to beneficiaries by means of authentication. The idea was that once we initiate biometric authentication using Aadhaar, it would bring in the desired transparency and ensure that only those who were entitled got the ration and nobody else could corner their entitlement. This would also bring in portability to enable people to withdraw their entitlement from anywhere in the country.

I was asked to present to the NAC, UIDAI's view on how Aadhaar could be used in the PDS. From the very beginning of my presentation, it became clear that some of the members had already made up their mind to contradict whatever I was trying to say. When I proposed biometric authentication at the point of delivery of foodgrains, they made me feel as if I had no idea what I was talking about. They asked me whether I was at all familiar with the PDS. When I informed them that I had dealt with the PDS in the field in Bihar for about a decade and had travelled to a number of places in the country and had seen biometric authentication being done in some districts of Andhra Pradesh and other places, they looked at me with contempt—as if I had no understanding of the situation. I felt insulted and knew that these people would oppose the project, as they thought it would harm the poor.

Sure enough, Roy and Drèze approached Mrs Sonia Gandhi with the request that the project be shelved outright or stopped till its validation by NAC. Nandan too made a presentation before the chairperson of NAC (Mrs Gandhi) and was successful in convincing her that shelving the project was not a great idea. If NAC had had its way, the project could have died a premature death. However, it survived.

Parliamentary Challenge: Among the strongest and indeed almost vitriolic of oppositions to the Aadhaar project came from the Parliamentary Standing Committee on Finance when it suggested that the project be jettisoned altogether. In its 2011 report, it suggested:

> The UID scheme has been conceptualized with no clarity of purpose and leaving many things to be sorted out during the course of its implementation; and is being implemented in a directionless way with a lot of confusion.

The National Identification Authority of India Bill was introduced in the Rajya Sabha on 3 December 2010. The Cabinet approved the

Bill that was then referred to the Standing Committee on Finance on 10 December 2010 for examination and report thereon.[164]

The first sitting of the committee took place on 11 February 2011. In that meeting, Nandan made the presentation on the UID project. The committee held its second meeting on 29 June and heard the views of the critics of the UID scheme. Its third meeting was held on 29 July 2011. The committee did not consider it necessary to provide an opportunity to respond to the critics. The committee met on 8 December 2011 and adopted the report and the same was submitted to the Lok Sabha. There were three dissenting notes to the report—from Members of Parliament (MPs) Raashid Alvi, Prem Das Rai and Manicka Tagore—largely on party lines.

The committee completely and unambiguously trashed the UID project and made some scathing observations. It concluded that the Bill in its current form is unacceptable and also recommended the closure of the scheme. It is interesting that the committee did not even consider it necessary to go into the provisions of the proposed Bill. The entire report focussed on the implementation of the project—its technology, conceptualization, interdepartmental issues and other similar areas. On receiving this report, we were more or less certain that the project will die a sudden death. Yet, it survived.

Acceptability Test: While Aadhaar was being created, we started looking for its usage. We created a number of papers such as financial inclusion,[165] Aadhaar and telecom, Aadhaar and health, DBT, etc. The major challenge was to make these systems accept Aadhaar as proof of identity and proof of address, which each domain requires at the point of access and entry. Similarly, states too required these proofs of identity and address for their schemes.

While it appears natural that Aadhaar be accepted as an ID

---

[164]The Bill was referred to the Standing Committee on Finance by the Speaker, Lok Sabha under Rule 331E of the Rules of Procedure and Conduct of Business in the Lok Sabha.
[165]'From Exclusion to Inclusion With Micropayments,' Unique Identification Authority of India, Planning Commission, April 2010. Available at: http://indiamicrofinance.com/wp-content/uploads/2010/12/Exclusion_to_Inclusion_with_Micropayments.pdf. Last accessed on 8 May 2020.

proof everywhere, it was difficult to convince departments in the early stages. The usual criticism was that the process of Aadhaar enrolment was weak and in private hands and hence, the data was not credible. One could easily change the name and address and there was no good process for verification of ID and address. Later, some domains admitted that Aadhaar could be accepted as proof of identity, as the loose process of identification could be taken care of by a robust process of deduplication, but it was not possible to accept it as an address proof, as no amount of technology at the back end could correct the problem. While we argued that in this connected world, proof of address was becoming less relevant, many domains whose systems were based on the place where a resident is residing, did not accept this argument.

We had to carry out a long and sustained struggle to get Aadhaar accepted both as proof of identity and proof of address. We also had to struggle to get the Aadhaar authentication accepted at the point of service delivery. Similar is the story of eKYC. However, once different domains realized the ease of using Aadhaar and the robustness of authentication/eKYC, there was no stopping.

Section III

AADHAAR COMES ALIVE

Chapter 9

PROOF OF THE PUDDING

Anything that won't sell, I don't want to invent. Its sale is proof of utility, and utility is success.

—Thomas A. Edison

The Immigration & Checkpoints Authority (ICA) of Singapore verifies the passport and other paperwork of any foreigner entering the country. Once confirmed, the details are linked to the visitor's thumbprints. Besides addressing Singapore's security concerns, the system allows the visitor to leave the country by scanning their passport and thumbprint with a device connected to the exit gate. No other human intervention is needed. This is an example of linking and storing biometrics for a single purpose. Other systems could profitably take similar steps.

Aadhaar too enables everyone to implement such systems easily. It links the identity of an individual to a randomly allotted 12-digit number and stores it with a hash of the fingerprint minutiae.[166] Systems that need to verify a user's identity can now build the functionality of biometric authentication in half a day: all the heavy-lifting to scan, process and store the fingerprints of more than a billion residents having already been done by UIDAI. Further, UIDAI's deduplication process before allotment of the Aadhaar number ensures that one individual cannot create two identities in the system.

Aadhaar thus provides the authentication to verify identity, bringing in the technology to support systems trying to bring in efficiency, while banishing imposters and ghosts. Authentication

---

[166]Fingerprint Minutiae, Webopedia. Available at: https://www.webopedia.com/TERM/M/minutiae.html. Last accessed on 15 June 2020.

devices can be in any form—mobile phones, tablets or PCs. Service providers can perform the authentication so long as they follow the open standards set by UIDAI. The communication mode can be anything, ranging from broadband to simple GPRS, and the time taken for authentication is, at the most, a couple of seconds. And the cherry on the cake? UIDAI provides this service free of cost as a public utility, making it virtually costless to adopt. The proof of the pudding, however, is in the eating. Here are some sample tastings, of which I have had first-hand experience.

Marking Biometric Attendance

As Chief Secretary in my home state of Jharkhand, after leaving UIDAI in March 2013, I was keen to try Aadhaar for reforms in implementation of the PDS, NSAP, MGNREGA, scholarships and other areas.

When I tried to explain to the officers in the state what Aadhaar's authentication service can achieve, it drew a blank. So, I thought of demonstrating it to them in their own offices and in a manner that made the impact immediately clear. What could be better than to have an Aadhaar Enabled Biometric Attendance System (AEBAS) in the government? We had already tried it in the UIDAI office in Delhi for our own staff on an experimental basis and I was convinced that it could be easily scaled up.

It was the time of President's Rule in Jharkhand. I convinced the Advisors to the Governor and the Governor himself about the urgent need to introduce the biometric attendance system, as there were serious problems of absenteeism, late arrivals and early departures, which demotivated employees who were sincere and punctual. Everybody acknowledged there was the need to improve the work culture, starting with regular and punctual attendance.

With AEBAS, people keyed in a small number (to which their Aadhaar was linked) and touched a finger to the scanner. Almost instantly, UIDAI would confirm their identity, which constituted proof of presence for officers and staff at the start and close of working hours.

Punching clocks have performed a similar function in factories. But the impact of AEBAS was fairly dramatic. First, the attendance could not be marked by proxy. Secondly, you did not require any card. Cards can be left at home. Not your fingers! Not only was the system robust, it did not require additional manpower and the attendance was tallied as it was marked.[167] So, it was possible to know how many persons had joined office in real time!

Both regularity and punctuality in attendance improved as a consequence, though I did become unpopular among government officials. Others joked that I had turned them into *angootha-chhaaps*—or illiterates who couldn't sign their name, so must affix a thumb impression. Eventually, however, improved attendance led to the switchover from a six-day to a five-day work week at the Secretariat. The system was later extended to district and block offices, and even to schools and hospitals in mofussil areas.

The PM Wants Biometric Attendance in Delhi: Following elections in the country, the new government had taken charge at the Centre in May 2014, when I joined as Secretary in DeitY. As mentioned earlier, during the departmental review with the PM, he enquired whether the attendance system I had implemented in Jharkhand could be introduced in Delhi. I answered in the affirmative and when asked for the timeline, I promised to start the system by 1 October 2014, i.e., in just over three months. I have written about it in the previous chapter.

We reworked the software from Jharkhand. We had developed this system on an open-source stack, which is inexpensive, robust and scalable. The team that completed both the server-side and client-side software comprised only three people—two being those who did it in Ranchi. The only cost was the procurement and installation of devices at the front end and their maintenance. This work was done by NIC. The trial started on 1 October 2014, as committed.

As the connectivity requirements of this system are met even

---

[167]Jharkhand government's attendance dashboard. Available at: http://attendance.jharkhand.gov.in/. Last accessed on 8 May 2020.

with low bandwidth, it was possible to extend it to any office or place with mobile coverage, including government offices, schools, health centres, anganwadis and midday meal centres. At a minimum, the system replaced the manual attendance registers with digital registers, allowing analytics and processing of data for decision-making. Other advantages are improvement in system efficiency and reduction of leakages.

Today, more than three million employees in various government offices of the country—starting from the Central Secretariat and down to the block level—are using the biometric attendance system. It has spread to district offices, universities, colleges, hospitals, village panchayats and municipalities everywhere. Yet, many people do not know that the system is based on the biometric authentication service provided by Aadhaar.[168]

Aadhaar-Based Systems Are Simpler

Biometric attendance systems are not new and have been tried at different points in time in several ministries and departments. However, all these systems quickly withered away. How was the present system more sturdy for survival then?

The answer is simple. In closed systems, the entity seeking to use a biometric attendance system has to build, keep safe and operate all biometrics-related functions, such as registration, deduplication, matching and life-cycle management of biometrics of all employees. This makes the system complex and costly. It also creates a dependence on the original vendor to maintain a customized, closed and often non-scalable solution. Add to that the need to maintain inventory of specific front-end devices that are part of the attendance system. All in all, it made the system too unwieldy.

On the other hand, AEBAS has no biometric components at all. It leverages a common authentication infrastructure established by UIDAI that has the capability of serving millions of authentication requests per hour, with a response time under two seconds. The

---

[168]The reader may visit one of the attendance portal at attendance.gov.in.

infrastructure, which can scale to keep pace with the aggregate all-India demand, is UIDAI's headache. Adoption of open standards and open architecture makes the system interoperable, extensible and allows commodity tablets, PCs and fingerprint or iris sensors to be used as front-end devices. These factors keep the system lightweight, inexpensive and easy to maintain.

There are other benefits too. For instance, an employee can mark attendance at any device anywhere in the system, irrespective of the department or offices. Therefore, a government officer from Delhi, while on official duty in an office in Mumbai, could mark attendance from the office being visited.

Lastly, the system provides insights and real-time analytical information on a dashboard. Data is generated because of activities and is not an activity itself, which is the case in many applications. Big data analytics on this data can help decision-making.

Digital India and Innovations around Aadhaar

Before I delve into other applications based on Aadhaar, I must talk about the Digital India programme of the Government.

Early in July 2014, BVR Subrahmanyam (Subbu), Joint Secretary in PMO, called to inform that the PM would like to launch a Digital India programme and our inputs were required. In a couple of days we compiled and sent them a list of e-governance programmes that could be taken up in the following months.

The PMO, however, wasn't looking for a few projects to launch. They said the PM's concept of Digital India was a comprehensive programme covering all aspects of digital space. That is, what kind of digital interventions would be required in India to transform the country into a digitally empowered society and a knowledge economy? They wanted us to develop a vision unconstrained by budgetary limitations.

So, we started thinking from scratch. With the help of Subbu and colleagues in the IT department, we designed a comprehensive programme and presented it to the Union Cabinet on 11 August 2014. The PM took personal interest in the programme. We created an overarching design with three main areas: Digital Connectivity as a utility to the citizens, Software and Services on Demand, and Digital Empowerment of the citizens. We also created measurable pillars of the programme. The experience of designing and articulating various components of the programme gave me a lot of interesting ideas, which became the drivers for several innovations on top of Aadhaar.

I must say that Digital India remains the PM's favourite programme till date and he takes personal interest in any initiative as he appreciates the role which technology can play in governance.

Jeevan Pramaan: Outcome of *Chai Pe Charcha*

On 1 November 2014, which was a Saturday, the PM had invited all GoI Secretaries for tea at his official residence at 7, Race Course Road. There was no agenda for the meeting, but many had carried departmental briefs as a precautionary measure. I was there as Secretary of DeitY.

The PM arrived and remarked that this was to be a free-flowing conversation. People eased up in a while and started relating personal anecdotes or sharing ideas. There was some discussion on more serious matters too, although in an informal way. For some reason, a discussion about pensioners and their problems started and I remarked to the PM that the requirement to provide a certificate that the pensioner was still alive (i.e., a life certificate), typically in the month of November, was a major problem. Pensioners had to do this personally, by presenting themselves before the bank manager. It was extremely inconvenient for elderly pensioners living away from the disbursing branch or unable to travel due to illness. I said, much like biometric attendance, a system could be developed to allow biometric authentication of pensioners living anywhere in the world to prove that they were alive. The PM nodded that this was a good idea and the conversation moved to other matters. But the idea had taken hold of me.

I could barely wait for the weekend to get over to start work on it. As the life certificates typically are submitted in the month of November itself, we realized that if we did not develop and deploy a system in the next few days, we will have to wait for the next year.

We worked at a frenetic pace: we developed the software, completed the documentation and named the facility Jeevan Pramaan, meaning 'life certificate'. We printed a single-page brochure and tested the software—all in a matter of three working days!

The PM was to leave on Tuesday, 11 November, for the 25th India–ASEAN Summit and East Asia Summit being held in Myanmar. Thereafter, he would attend the G20 Summit in Brisbane, Australia, and meet leaders of the Pacific Islands at Fiji.

I told his office that if we didn't inaugurate the system on Monday itself, pensioners won't be able to use it that year. Luckily, we got the time. We informed and invited everybody—our own minister, the PMO, the Cabinet Secretary, the Expenditure department and the Pension department—for the launch at 4 p.m. on 10 November 2014. Everything went off smoothly and the PM remarked: '*Dekhiye, chai pe charcha ka kuchh to fayada hua na!* (Look, something positive came out of a chat over tea!)'

Of course, the government doesn't work like this. Projects don't get implemented and launched in this manner, and that too by the PM! I was duly reprimanded for that. However, the programme for Digital Life Certificate (called Jeevan Pramaan) was launched within a week from its conceptualization. Today, it is being used by millions of pensioners.[169] How could it conceivably happen so fast? Simply put, the required components had already been built by Aadhaar! Integration becomes easy once you have well-defined and open APIs.

ORS at Hospitals

One morning in 2015, I was sitting with V. Srinivas, the deputy director (Administration) of the All India Institute of Medical Sciences (AIIMS) at New Delhi. I was there to seek his help in admitting a patient.

---

[169]Jeevan Pramaan—Govt of India. Available at: https://jeevanpramaan.gov.in. Last accessed on 8 May 2020.

There was a huge crowd at the counters for appointments and admissions. The deputy director told me that patients had to wait for a long time and the outstation ones often camped with their families around AIIMS. The hospital had only limited slots for outpatient department (OPD) appointments in each speciality. Patients queued up since early in the morning for the limited number of slots. If they were unsuccessful, the only option was to join the queue, even earlier, the following day.

Where there is misery, there are middlemen to profit from it. They could help you jump the queue by purchasing bookings made in advance. The failure in instituting a smooth booking process caused other problems at the hospital. For instance, if patients did not turn up at the OPD clinic after securing the appointment, it was a loss to everyone. So, we discussed if technology could ease this misery. I returned from the hospital, somewhat overcome by the distress of those suffering from serious illness or dire poverty, the two situations that drive people to AIIMS.

What if patients could fix an appointment online and then show up on the appointed day? They could thus avoid the long queues and save money and inconvenience for themselves and their family. But for such a system to work, the problem of fake bookings and no-shows had to be addressed first.

The issue of fake bookings was resolved through OTP-based authentication sent to the Aadhaar holder's registered mobile number. Once the OTP is verified by UIDAI, the appointment is granted, thus ensuring that the booking cannot be subsequently transferred to another patient. In other words, appointments are a sequence of Aadhaar numbers, which uniquely identify the patient. And this sequence cannot be reordered because the booking time carries an authentication timestamp that could be used for an audit.

The problem of no-show would still exist. This problem can be mitigated by a well-designed system that rechecks the patient's intention closer to the time of the appointment and offers vacated slots to other people. This thinking led to the creation of the online registration system (ORS).

After a discussion with Srinivas, we created a small team of

NIC officers under Rajesh Gera and started working. We added other hospitals once the appointment system became operational. Because it was hosted on cloud infrastructure, it could scale up with need. Other facilities too were added in the workflow and finally ORS[170] was formally launched by the PM on 1 July 2015, among other DeitY projects. The list of hospitals brought onto this system keeps growing across the country and has touched 182 as of 2019 and provided more than three million appointments since its launch, till date.[171]

Digital Signatures on Demand (eSign)

The PM launched several other initiatives, besides ORS under Digital India in July 2015.[172] Two of the more important initiatives were eSign[173] and Digital Locker,[174] both essentially based on Aadhaar.

The idea of eSign came first. For those unfamiliar with digital signatures, it is a mathematical scheme or an electronic fingerprint that ensures that an electronic document or a digital message (email, spreadsheet, text file, etc.) is authentic. A valid digital signature tells the recipient that the message was created by a known sender—who is the person he claims to be—(authenticity), who cannot deny having sent the message (non-repudiation) and confirms that the message was not altered in transit (integrity).[175]

Based on rigour and the purpose, digital signatures are divided into three classes. The highest class—class 3—is the most trusted

[170]The ORS portal is located at http://ors.gov.in/copp/. Last accessed on 8 May 2020.

[171]ORS for Listed Hospitals. Available at: https://ors.gov.in/copp/more.jsp. Last accessed on 8 May 2020.

[172]The Launch of Digital India Week by PM Narendra Modi, Doordarshan Channel. Available at: https://www.youtube.com/watch?reload=9&v=NPHZeoui4QA. Last accessed on 9 May 2020.

[173]eSign–Online Electronic Signature Service, Website: Controller of Certifying Authorities (CCA), Ministry of Electronic & Information Technology. Available at: http://cca.gov.in/eSign.html. Last accessed on 9 May 2020.

[174]DigiLocker: Document Wallet to Empower Citizens. Available at: https://digilocker.gov.in. Last accessed on 9 May 2020.

[175]Digital Signature. Available at: https://en.wikipedia.org/wiki/Digital_signature. Last accessed on 19 May 2020.

digital signature class recognized by the Information Technology Act, 2000, and with the same legal validity as an ink signature. Promoting the use of digital signatures is one of the functions of the Department of Information Technology, GoI.

A digital signature certificate (DSC) is a secure digital key issued by the certifying authorities for the purpose of validating and certifying the identity of the person holding this certificate. Digital signatures that make use of public key encryptions to create the signatures may be issued by the certifying authorities.[176]

The process of issuance of this class of digital signature includes the ID verification of the individual seeking the DSC. After ID verification, which usually requires the physical presence of the person, the keys used for creating the electronic signature are stored in a hardware cryptographic token secured with a password/PIN number and delivered to the holder in the form of a USB dongle. The holder of the DSC can now sign any document by plugging in the USB in his computer and using the PIN.

The cost of this dongle (including the ID verification cost) comes to about ₹1,100. In most cases, a person rarely needs DSC; perhaps once in a year at the time of filing income tax returns. The currency of the DSC is two years. Hence, essentially a person spends ₹550 for one digital signature! The relatively high cost and the need to reissue the USB dongle every two years have prevented digital signatures from being widely used in India—we had only seven million DSCs in the whole country, largely with government officials for signing various types of certificates (caste, income and residence certificates, whenever these are issued online).

Simple Process; Lower Cost: We introduced online digital signature or eSign that helped to ease the process and reduce costs. Clearly, issuing a DSC required identity verification and issue of the DSC dongle. We used the online eKYC process that UIDAI had already

---

[176]Information Technology (Certifying Authorities) Rules, 2000 lay down detailed guidelines relating to Digital Signatures. Rules for Information Technology Act 2000. Available at: https://meity.gov.in/content/rules-information-technology-act-2000. Last accessed on 9 May 2020.

created as an extension of its authentication services, on the basis of which we could issue a digitally signed copy of the person's Aadhaar.[177] The eKYC is the virtual world equivalent of the ID document in the physical world to be used by the digital certificate provider or the certifying authority, who issues the DSC dongle.

Once the first process became real time and online, I wondered if it wasn't possible to similarly treat the second. After all, issuing a Digital Signature is essentially creating a pair of keys for asymmetric cryptography. Rather than storing the key pair on the dongle, could we not release them online at the time of signing a document? This would provide the digital signature facility to anybody with an Aadhaar number. What's more, each time a person had to sign, a new key pair could be generated, providing more robust key management. In this scenario, the job is effectively unbundled in two parts: the first job of ID authentication and eKYC issuance is done by UIDAI and the second work of issuance of one-time DSC is done by the Certifying Authority (CA).

We called this online digital signature as eSign and the entity using this process to issue digital signatures as the eSign provider. eSign made the major cost elements of the earlier issuance process such as ID verification, maintaining ID records and the cost of the USB dongle disappear! We also developed design documents and carried out appropriate modifications in the rules to provide for entities that could provide these digital signatures.

I am told that today there is a healthy ecosystem of eSign providers and it has started being used in various applications. Most importantly, anybody who has Aadhaar is capable of digitally signing any document and this facility is available on demand on a real-time basis. There are both public and private entities providing eSign facilities.[178]

---

[177]Aadhaar eKYC. Available at: http://www.e-mudhra.com/aadhaar-ekyc.html. Last accessed on 9 May 2020.

[178]Controller of Certifying Authorities, Ministry of Electronics & Information Technology, Govt of India. Available at: http://cca.gov.in/about.html. Last accessed on 1 June 2020.

Digital Locker: Private Space on Public Cloud

On 30 March 2019, Sudhir Prasad, ex-chief secretary of Jharkhand, sent me a message. His car had been involved in a minor accident and for making the insurance claim he needed the Registration Certificate (RC) of the car. Poor Prasad had misplaced his RC. The District Transport Officer at Ranchi was unable to issue a duplicate because the original document had been sent to the record room and was difficult to retrieve. However, Mr Prasad was delighted that he was able to locate the RC in his Digital Locker and sent a message thanking me.

The concept of Digital Locker was articulated in the Digital India vision.[179] It was meant to be a utility to citizens and promised 'shareable private space on a public cloud' for every citizen. The programme called it the Digital Locker. At the time of articulating this programme and its components, developed in early 2015, we did not have the concept of eSign. Hence, we had merely thought of Digital Locker as a digital storage, something like Google Drive or Dropbox.

However, once the idea of eSign was concretized, we introduced the concept of digitally signed documents being stored and shared rather than just issuing digital documents. eSign facility was attached with the digital locker making it more versatile and acceptable. We also developed the concept of digital locker service providers, document issuers and document depositories. Hence, Digital Locker is a private space provided by the government where Aadhaar number is the ownership identifier.

An ecosystem of entities that issues and accepts digitally signed documents stored in the citizen's digital locker is growing. Entities such as the Central Board of Secondary Education (CBSE) or other education boards, and regional transport officers are able to push digital certificates in the digital lockers of citizens. Fake degrees and certificates could soon become a thing of the past.

---

[179]Digital India. Available at: https://www.digitalindia.gov.in/content/vision-and-vision-areas. Last accessed on 9 May 2020.

Today, the DigiLocker[180] portal has more than 33 million users storing 3.7 billion digitally signed documents[181] that have been issued by various authorities. More than 300 million educational certificates are available on DigiLocker. All new vehicle RCs and driving licences are pushed into the DigiLocker of the Aadhaar holder. You do not need to carry these documents because you can show them directly from your DigiLocker, accessed through your mobile. Even Indian Railways now accepts digital Aadhaar and driving licence from DigiLocker as valid ID proof.

Your ID and other important documents are always with you in your mobile, safe and digitally signed. Even if you do not use your account, your account is there, and you can use it anytime in future. Products developed with Aadhaar that incorporate Aadhaar's philosophy are changing the way people live, be it convenience in paying parking charges or dealing with life's rarer events that one insures for.

I feel fortunate to have been involved not only in implementing Aadhaar but also in conceptualizing, developing and implementing some of these important applications based on this identity infrastructure. The fact that these systems have brought a positive change in the lives of hundreds of millions of people is a source of great personal satisfaction.

---

[180]Digital Locker: Document Wallet to Empower Citizens. Available at: https://digilocker. gov.in. Last accessed on 9 May 2020.
[181]DigiLocker. Available at: https://en.wikipedia.org/wiki/DigiLocker. Last accessed on 19 May 2020.

Chapter 10

CONVINCING THE PARTNERS

A garden requires patient labour and attention. Plants do not grow merely to satisfy ambitions or to fulfil good intentions. They thrive because someone expended effort on them.

—Liberty Hyde Bailey, American horticulturist and botanist

You want to remodel your garden and have the means to pay the gardener? Go ahead. You may rue the decision if the job takes longer than anticipated or costs more than your estimate, but at least it can be done. If you want your neighbour's garden to be remodelled, it's considerably more problematic even if you pay the gardener and buy the required items. However, the trickiest remodelling is when the garden and the expense belong to the neighbour and you have only ideas to implement. Your only hope in such a case is to find a champion across the fence to own those ideas! In the government, IT projects come in all three flavours and you can imagine how they progress according to type. The UID project belongs to the last tricky category.

All that UIDAI offered was a number to the residents and the idea to government departments that they could seed the number in their databases and thus ensure that a resident figured only once in their list. Government departments could also identify the resident at the point of service delivery through a biometric scan or an OTP sent to their registered mobile number.

It was the allies we won within and outside the government who became champions of the project in their organizations. Sometimes, it was sheer timing that saved the day even as the project looked doomed.

Financial Inclusion

With his vast experience in banking, Nandan was of the view that one of the first and most compelling use for Aadhaar was in financial inclusion and payment systems. As early as April 2010, a full six months before the issuance of the first Aadhaar number, UIDAI brought out a booklet titled *From Exclusion to Inclusion with Micropayments*.[182] This booklet provided an architecture for facilitating the opening of bank accounts using Aadhaar for KYC and supporting low-cost digital transaction services through business correspondents, using Aadhaar authentication on Micro ATMs.

The proposal asserted that the new architecture will reduce the cost of banking and hence, make it viable for banks to provide low-value but high-volume services to the poor. The proposed architecture assumed availability of mobile connectivity to carry out remote authentication and banking transactions.

Officers in the Department of Financial Services (DFS) were aghast at our proposal. How could an outsider with neither the mandate nor expertise even utter the word 'financial inclusion'? It was their garden to tend and they would brook no unwarranted interference. The more we tried to convince the DFS by selling our 'product', the more vehement was the pushback.

Turf issues are common in the government. In this case, however, the intensity of resistance was extraordinary. UIDAI, we deduced, was being viewed as a private outfit trying to influence government policy. Though he was Chairman, UIDAI, Nandan was the poster boy of private entrepreneurship in India. Moreover, as UIDAI had people from both the private sector and on deputation from government departments, the organization seemed to lack legitimacy in the eyes of the dyed-in-the-wool bureaucrats.

The enthusiasm of our team in selling Aadhaar didn't go down well either. The departments would question the 'product' critically and conclude that it didn't meet their requirements, without taking

---

[182]*From Exclusion to Inclusion with Micropayments*, UIDAI, Planning Commission, April 2010. Available at: http://indiamicrofinance.com/wp-content/uploads/2010/12/ Exclusion_to_Inclusion_with_Micropayments.pdf. Last accessed on 9 May 2020.

the time to understand the nuances. They seemed wary of the eager salesman.

The secretary of one such department told Ashok Pal Singh, then DDG, looking after Financial Inclusion in UIDAI, that while digital technologies could indeed make banking more inclusive and viable, he would develop his own technology and not use what UIDAI had to offer. When Ashok informed him that UIDAI's technology was also his government's technology, the secretary did not agree. He tried to develop an alternative way to implement the DBT scheme that eventually failed. Later, Aadhaar, as a universal financial address, proved to be a game changer for implementing DBT.[183]

For every solution, there was a problem—or so it seemed. The DFS refused to be drawn into specifics. Financial inclusion and opening bank accounts for the unbanked was their mandate and they were capable of taking care of it. The banks took their cue and raised objections. The rulebook is the last refuge of the bureaucrats. Aadhaar, they said, was not a valid document for bank accounts under the rules of the Prevention of Money Laundering Act, 2002.

We, at UIDAI, were equally determined in the belief that for every problem, there ought to be a solution. So we trotted off to the Department of Revenue that notified documents for opening bank accounts. We hit a wall in trying to explain that a virtual ID that is not in the form of a card could, nonetheless, exist. We persisted and when it seemed that our arguments could not be refuted by logic alone, a bureaucratic googly was bowled to us.

The RBI was ushered in as the sector regulator, in the belief that its innate conservatism would prevail. To compound the issue, a reference was also made to the Financial Intelligence Unit (FIU), MoF, to examine the feasibility of Aadhaar as a valid document for financial transactions.

Fortune may not favour the brave, but it does favour the persistent. The tide unexpectedly turned in favour of UIDAI when FIU, though not convinced about the workability of a virtual ID and

---

[183]R.S. Sharma, 'Aadhaar Bill: Click out the missing link,' *The Economic Times*, 5 March 2016. Available at: https://economictimes.indiatimes.com/blogs/et-commentary/aadhaar-bill-click-out-the-missing-link/?upcache=2. Last accessed on 9 May 2020.

online authentication, got excited about a paperless process to share ID and address information, server to server. It led to the invention of eKYC, where an Aadhaar holder can authorize UIDAI through biometric authentication to share a photograph and address with a service provider, securely in real time. Realizing that this was the most secure way to undertake KYC, FIU approved it. R.S. Gujral, then revenue secretary, also got the import of it, and in the next meeting, pinned both the RBI and the DFS to either accept this solution or argue their way out of it. With nothing else to say except that it was an untried solution anywhere in the world, the RBI and the DFS signed on the dotted line.

The commercial banks subscribed willy-nilly to Aadhaar for KYC, once it was notified by the Department of Revenue as a valid document for financial transactions. Visa saw an opportunity there and, working with three banks, did a pilot for instantaneously issuing prepaid cards against Aadhaar eKYC. This initiative allowed UIDAI and the banks to test the product, although it took longer for wider adoption. Once the new government came to power in 2014 and the PM gave a new lease of life to Aadhaar, the adoption process got accelerated. Soon, the entire IndiaStack—a popular name for various digital artefacts developed on top of Aadhaar and using its broad architectural principles—utilizing the digital infrastructure for paperless and cashless service delivery came to be built on top of it.

Working with the National Payments Corporation of India (NPCI),[184] UIDAI built the Aadhaar Payment Bridge (APB)[185] to channelize government benefits and subsidies to rightful hands. AEPS enabled benefits' withdrawal only by the genuine beneficiary, eliminating proxies and bogus beneficiaries. Next, eKYC spread its wings beyond banking to related subsectors in the financial sector, allowing user agencies, public or private, to verify their customers

---

[184]The National Payments Corporation of India (NPCI) is an umbrella organization for operating retail payments and settlement systems in India. To know more, visit: https://www.npci.org.in/

[185]'How Aadhaar Payment Bridge (APB) System Works?' Available at: https://www.npci.org.in/how-aadhaar-payment-bridge-apb-system-works-0. Last accessed on 9 May 2020.

online using biometric scan or OTP over Aadhaar-linked mobile number.

It's in their DNA to be conservative, but eventually banks and the financial sector enthusiastically adopted the Aadhaar-based ecosystem to their great advantage. Not just the RBI but all financial-sector regulators—Insurance Regulatory and Development Authority (IRDA), Pension Fund Regulatory and Development Authority (PFRDA) and Securities and Exchange Board of India (SEBI)—adopted Aadhaar soon enough.

Catching the Fancy of Telcos

It took several years for the system to accept that Aadhaar could be a sufficient KYC for getting mobile phone SIMs. In 2010, UIDAI published a monograph titled *Aadhaar Enabled Applications in Telecom*, which proposed that Aadhaar authentication 'offers a solution that can deliver inclusive access to telecom services without compromising on subscriber verification.'

The proposal to use Aadhaar authentication/eKYC to provide SIMs to people did not get the go-ahead from the Department of Telecommunications (DoT) for many years. Although the DoT was in favour, the proposal continued to be vetoed by the MHA for security reasons—perhaps due to turf issues between the RGI and UIDAI. Their refrain was that Aadhaar-based KYC for mobile connections would compromise security because it was not a robust proof of address required for granting mobile SIMs.

eKYC first caught the fancy of the telcos struggling to comply with KYC norms following a paper-based process. It was well known that SIM cards were sometimes issued without KYC. Furthermore, photocopies of documents left by a new subscriber were copied by unscrupulous agents to issue SIM cards to other customers without KYC documents. The telcos were incurring huge costs in collecting physical copies of KYC documents from their agents and storing them for long periods in godowns across the country. They had accumulated fines amounting to thousands of crores of rupees due to their failure to produce KYC documents. Customers, too, were dissatisfied not only due to the misuse of their ID documents, but also due to delays in the activation of their SIM cards while awaiting authentication of their documents.

The Cellular Operators Association of India (COAI)[186] asked UIDAI for a demo-cum-presentation in which the DoT also participated. For COAI, it seemed like a dream solution, as eKYC could do away with presentation and storage of physical documents and make it impossible for an agent to either issue a SIM without KYC or fraudulently reuse documents of one customer for another. Besides, a secure real-time server-to-server transfer of KYC data from UIDAI to the telco-enabled instant activation. This potential was to later become the default method of applying for a Jio SIM connection and its record-breaking customer acquisition speed.

COAI endorsed eKYC enthusiastically, but the DoT was wary of using an out-of-the-box solution without precedence, given the security implications of mobile telephony. It promptly referred its intention to allow eKYC for issue of SIM cards to the MHA.

It was one thing for COAI and the DoT, both familiar with information technology, to grasp the usefulness and security of eKYC, but quite another for the MHA to accept the arguments. They were determined to thwart UIDAI, who they viewed as johnny-come-lately, to encroach upon their turf. They brought in the Intelligence Bureau (IB) to weigh in for them and soon, the arrival of a new Home Ministry brought matters to a standstill. They upped the ante, extending the security argument to whole of Aadhaar, and sought for it to be merged with their own NPR exercise.

An interesting facet in turf protection is that participants tend to switch sides with roles. So, it happened when a senior officer of UIDAI, who had occupied a key post and batted strongly in favour of Aadhaar for years, did a somersault the moment he was posted in the MHA. Quite remarkably, he took to spitting venom against UIDAI's enrolment process and 'exposing its underbelly'. This

---

[186]The Cellular Operators Association of India (COAI) is an industry association of mobile service providers in India.

phenomenon affected not just the bureaucracy, but also the political executive. Much to our delight, we found that Mr Chidambaram, who had earlier opposed Aadhaar when he was the home minister, championed it for its DBT capabilities when he became the finance minister.

UIDAI continued to try and convince the DoT and the MHA that in the context of mobile connections—especially when 95 per cent of mobile connections are prepaid and granted without address verification—using Aadhaar KYC will be more robust than the existing system of paper documents, both from the security and customer point of view. The MHA frequently bandied the bogey of Aadhaar being misused by refugees and terrorists, much like their early opposition to mobile telephony being used by criminals. Over time, however, mobile phones have become a major aid for tracking crime and criminals.

In the meantime, in its communication to the DoT in December 2014, TRAI had suggested a move towards Aadhaar-based eKYC. Considering developments in the intervening period, TRAI reiterated the same recommendation in January 2016 with slight modifications. Finally, in August 2016, the DoT permitted Aadhaar-based eKYC as an alternative process for acquiring new subscribers.

Linking ID to Bank Account

Although cash transfers were not unknown prior to Aadhaar, linking an ID to a bank account was pioneered by UIDAI. This again was not a smooth affair. For instance, the MGNREGA programme and NSAP were two central schemes widely in operation across states where cash benefits were provided. In both the cases, prior decisions existed for payment through bank accounts.

When UIDAI architected APB with NPCI, it made Aadhaar the payment address of the individual by linking the Aadhaar number to the bank account number of the beneficiary. APB effectively allows cash transfer merely against the Aadhaar number without quoting the bank account. Here again, we came up against an unexpected wall.

The prevalent model was that each government department would sign up with a bank of its choice to transfer cash benefits. Thus, MGNREGA wages came in one account, NSAP into another and maternity benefit into a third account. The idea of a single bank account linked to Aadhaar meant loss of accounts for banks. More importantly, an individual with many accounts could choose the bank account to be linked to Aadhaar, weakening the captive arrangements devised by officials and bank managers.

The DFS opposed APB simply as something beyond the remit of UIDAI. Bank managers goaded state governments to pose problems. All three combined to undo APB by insisting that there should be a provision to link multiple accounts to Aadhaar on the specious plea that it was the right of an individual to receive different benefits in different accounts!

The reality was that social security recipients were inconvenienced by having to maintain multiple accounts. The splitting of bank accounts detracted from economies of scale for any bank to deploy business correspondents to serve the poor and the unbanked. Government departments found the task of obtaining bank account details and transferring cash benefits herculean. In the absence of a unique identifier and multiple bank accounts, there was no way to separate the fakes and duplicates in the system. Aadhaar, through APB, solved all these problems. Once again, persistence and sound design principles paved the way for acceptance of APB linkage to a single account with the help of a few champions in the states who saw merit in UIDAI's work.

States Were Better!

State governments were relatively easy customers both for propagation of Aadhaar as well as various Aadhaar-enabled systems such as APB. As they had been involved as Registrars to UIDAI, they had first-hand knowledge about the adequacy of safeguards in the enrolment process. They also knew that the UIDAI back-end systems were robust enough to accommodate simplifications in enrolment, which enabled the process to run smoothly.

This was contrary to the conventional way of doing things that involved demanding and stringent requirements, such as additional documents or a more involved verification procedure, which usually caused harassment, and worse, exclusion. UIDAI, on the other hand, made things easier at the front end by handling all complexities at the back end. Nandan would say that if you managed to enrol yourself as Rajesh Khanna, you'd have to be Rajesh Khanna for the rest of your life because the system would refuse to issue you another Aadhaar in any other name. The states realized this and so did the residents.

Secondly, the states also needed a person's ID in many of their programmes and had already started accepting Aadhaar in such programmes. We used to publish the notifications issued by various departments as and when they declared Aadhaar as valid proof of identity and proof of address in their domains. The purpose of such publications was to encourage other states and departments to also accept Aadhaar in their programmes. Ready endorsement by the states helped Aadhaar gain traction, both at the enrolment stage as well as its application.

Historically, the states were implementing GoI schemes, eagerly or otherwise, with diligence. They perceived Aadhaar as yet another GoI scheme and took it onboard. Even so, there would be some wrinkles to straighten out. The states assumed that issuing an identity was a sovereign function and so their exclusive preserve. It took them time to reconcile with the existence of non-state government registrars in this space, such as the banks and even the GoI's own Post Office.

The Life Insurance Corporation of India (LIC) was the first non-state government registrar. It was keen to be the first to kick-start the enrolment process in West Bengal, more particularly starting at Jangipur, the constituency of the then finance minister, Pranab Mukherjee. UIDAI was however dismayed to find that the local authorities abruptly stopped the enrolment process and confiscated the equipment. Fortunately, a letter to the Chief Secretary explaining the design helped and West Bengal became the first state to issue SOPs for enrolment by private agencies, with accompanying instructions to district authorities to facilitate their work. The Finance Ministry did get to inaugurate Aadhaar enrolment by LIC in the minister's constituency. A similar pattern occurred in many other states, which dampened the enthusiasm of non-state government registrars, but the problem was never as intractable as at the Centre.

When it came to application of Aadhaar for government purposes, UIDAI was lucky to find an ally in the undivided state of Andhra Pradesh. Even though Andhra Pradesh was already working on a local biometric programme and cashless disbursement of subsidies, it became the first to endorse Aadhaar for rations, pensions and MGNREGA. Initially, the Information Technology department of Andhra Pradesh was averse to using Aadhaar, as it was seen to be disruptive to its own forays, which were then reaching the scale-up stage. However, Nandan's meeting with the CM of Andhra Pradesh, facilitated by Union Minister Jairam Ramesh, brought attention to the Aadhaar programme. The visit happened at the time the IT secretary of the state had changed, and the successor was less invested in the state's homegrown solution. Call it propitious timing, but the moment was seized by Neetu Prasad, the DM of East Godavari district, and her deputy, Babu Ahmad, to demonstrate the efficacy of Aadhaar for PDS/Pension/MGNREGA disbursement. What added to the impetus was the coming onboard of R. Subrahmanyam as the principal secretary, Rural Development, and Harpreet Singh as principal secretary, Civil Supplies. For UIDAI, it meant sprouting of champions at critical levels!

Successful adoption of Aadhaar, DBT and related payment systems by Andhra Pradesh helped to gradually convince the GoI to accept Aadhaar for cleansing its beneficiary list of ghosts and duplicates. Likewise, the idea of channelizing benefits as cash to Aadhaar-linked bank accounts also began to gather favour.

Learning from Failure

Finance Minister Shri Mukherjee, in his Budget Speech of 2011–12, announced a task force to work out the modalities for a new system of direct transfer of subsidies—using Aadhaar numbers for kerosene, LPG and fertilizers—to be headed by Nilekani. The task force designed a solution that was flexible enough to implement diverse subsidy disbursement models. Pilots were conducted based on the recommendations of the task force. Their report was accepted by the finance minister, but implementation was left to respective ministries.

Once again, real progress was uneven. The Ministry of Chemicals and Fertilizers never really came onboard and adopted the age-old bureaucratic tool of endless procrastination to avoid inconvenient programmes. They were also at odds due to vested interests of the fertilizer lobby, which has huge stakes in continuation of subsidies that distort the pricing and the entire supply and distribution chain.

The task force had devised a system of selling fertilizer at market price and real-time transfer of subsidies to Aadhaar-linked bank accounts using biometric-enabled POS terminals/Micro ATMs deployed at the retail outlets of fertilizer dealers. The reluctance of the Ministry of Chemicals and Fertilizers combined with the tepid response of banks to deploy Micro ATMs put the entire initiative on the back burner. Strategically, given that it was acquiring traction with state governments, UIDAI decided to not pursue this track actively.

The kerosene component was heavily dependent on the state governments, besides being politically sensitive. The DBT for kerosene subsidies was launched in Alwar by the Rajasthan state government in January 2013. However, it did not succeed and took the will out of the political system to experiment with DBT for kerosene, given that the next general elections were in sight. Significantly, in the Alwar experiment, Aadhaar was not used to identify the beneficiaries, nor was it used to open accounts for fund transfer. Yet, critics have often attributed its failure to Aadhaar!

UIDAI used the learnings from this failure to structure Aadhaar-based initiatives. The early successes came from Andhra Pradesh for MGNREGA, pensions and PDS besides the LPG subsidy project of the central government itself.

While good timing helped UIDAI in Andhra Pradesh, the turning political tide torpedoed one of its biggest successes in Delhi. The Ministry of Petroleum had decided to try out Aadhaar for LPG subsidy. A local champion emerged in the form of Neeraj Mittal, joint secretary, who piloted the DBT for LPG (DBTL) scheme in 291 districts in the country from 1 June 2013 in six phases. It covered nearly 100 million consumers with over 3,770 distributors across the three PSU oil marketing companies with the objective of efficient subsidy administration. An amount of ₹5,400 crore (₹54 billion) was successfully transferred to almost 30 million LPG consumers across the country. While preliminary results indicated that the scheme met its primary objective of curbing leakages in the distribution system, the shadow of forthcoming general elections made the political class apprehensive of its political ramifications. The scheme was put in abeyance and a committee headed by Professor S.G. Dhande, former director of IIT Kanpur, was set up to review the scheme and submit its report to the GoI after consultation with the stakeholders.

The committee evaluated the scheme and concluded that the DBTL scheme was successful in achieving its objectives, viz. reducing diversion, eliminating ghost/duplicate connections and improving LPG availability. It also recognized that the scheme promoted enhanced financial inclusion. Finally, the committee strongly recommended that the DBTL scheme should be reinstated.

However, the momentum was broken, and it took a new government that assumed office in May 2014 to take it ahead after rechristening it as PAHAL. In due course, PAHAL became a flagship programme, one which brought popular accolades from the public, including electoral dividends.

Attracting the Private Sector

The success of the DBTL scheme, even though it was put in abeyance, spurred several user agencies, including those in the private sector, to register with UIDAI as AUAs of Aadhaar. There was a sudden spurt, with the number of AUAs soon touching 200. It appeared that the innovation platform was coming of age. The global information services company, Experian, which is into data analytics for the financial services sector, put up a proposal to use

Aadhaar for credit rating. Muthoot Finance designed an initiative to disburse cash against gold jewellery using Aadhaar in under three minutes wherein the beneficiary could collect cash at an ATM using the Aadhaar number and authentication. Hero Cycles looked at a fully automated rent-a-cycle programme. The cycles to be parked at transportation hubs could be unlocked and paid for using Aadhaar with zero security deposit.[187] In a similar vein, an NGO framed a scheme of providing LED lamps to people using kerosene lamps against Aadhaar authentication with no deposits.

The development of Aadhaar and related biometric applications attracted the gaze of the research and development (R&D) community of the world. Nokia was the undisputed leader of mobile handsets and it evinced keen interest in developing handsets with biometric readers. Its Chennai manufacturing facility was tasked with producing what may have been the first biometric-enabled mobile phones. Working in conjunction with the UIDAI team, a prototype was developed and tested. Sadly, around this time, the fortunes of Nokia nosedived and the project lost steam. Since then, we have seen the incorporation of biometrics in mobile phones by the likes of Apple and Samsung, both fingerprint and facial. Aadhaar is a landmark scientific development and needs to be carried forward by India. If not, we will lose our pioneering status.

Two factors inhibited the use of Aadhaar by the private sector. One was the lack of political will around 2014 to push for its adoption in the face of ensuing elections. The second was the high-decibel campaign launched by a handful of activists against Aadhaar in the aftermath of the PILs filed before the Supreme Court. Together, they created uncertainty around the future of Aadhaar. This was tragic, as it nipped the platform of its innovation base just as it was coming of age. The subsequent bucketing of Aadhaar as a mere utility for disbursement of social security benefits (one of the consequences of the Supreme Court judgement on Aadhaar) has been self-limiting as far as development of Aadhaar-based

---

[187]In case of theft, the chance of conviction was very high because the identity of the thief was known. Of course, one could never steal a second cycle because the system won't allow it.

applications are concerned.

While the Suprem Court cleared Aadhaar and its authentication services of all objections levelled by the petitioners, it limited its usage to only government programmes. This may have been a legal compulsion, but it does not seem logical to restrict the use of a public good to the government alone.

For many, Aadhaar is their first identity document or the one they prefer to use. They should have the freedom to use it anywhere, so long as such use is of their own volition. For instance, there is no bar to using your passport as an identity proof while registering a sale deed, though the passport's primary purpose is to certify the identity and nationality of its holder for international travel.

Further, Aadhaar is a public good (as defined by economists) because the authentication capacity of UIDAI is virtually limitless and no extra expenditure is involved with its increased usage. That is, authentication is non-rivalrous in the sense that its usage by one doesn't preclude another also from using it.

The benefits to the private sector in terms of reduced cost of service delivery will ultimately be passed on to the consumers and the paperless and digital processes will benefit the economy. Hence, in my view, voluntary usage of Aadhaar should not have been limited to government social benefit programmes alone.

In an increasingly online world, a solution to identify and authenticate the person will have to be found. Currently, the position seems to be 'anything, but Aadhaar'. The net impact may well be that from being a pioneer of virtual identification, India becomes, as usual, an adopter of a Western variant!

The GoI has passed the Aadhaar and Other Laws (Amendment) Act, 2019,[188] which enables the use of Aadhaar on a voluntary basis by individuals for obtaining services not just from the government but also the private sector. This should unleash its full potential to transform delivery of services to the residents of India.

---

[188]The Aadhaar and Other Laws (Amendment) Act, 2019, No. 14 of 2019, 23 July 2019. Available at: https://uidai.gov.in/images/news/Amendment_Act_2019.pdf. Last accessed on 9 May 2020.

Chapter 11

JAM AND OTHER RECIPES

Ideas are like rabbits. You get a couple and learn how to handle them, and pretty soon you have a dozen.

—John Steinbeck, American author

Traditionally, people are identified in the formal system by attaching the name of the father or husband and the person's address to their names. Aadhaar changed this because it uses biometrics to distinguish a person from everyone else, making furnishing address proof less important.

Aadhaar thus recognized the right of every resident to unfettered mobility within the national boundaries, whereby benefits are attached to the identity of individuals, irrespective of their place of residence. For example, the state of Andhra Pradesh introduced the idea of portability of ration entitlements to any place in the state.[189] This helps migrants who have moved for employment, typically from small villages to the cities. Today, a beneficiary can go to any ration shop and get his entitlement.

Transparency in PDS Transactions

The PDS is the largest distribution network in the world mandated to provide foodgrains and other items (ration) at subsidized rates. The system is administered by about 500,000 Fair Price Shops (FPSs) throughout the country. One of the complaints about the system is that the FPS dealers sell the ration items in the black market and show it as having been distributed to the people. FPS dealers deny

---

[189]AP PDS Portability. Available at: https://aepos.ap.gov.in/ePos/. Last accessed on 9 May 2020.

this allegation, while people claim that they were turned back on one pretext or the other. It becomes difficult to objectively investigate these claims and counterclaims, and corruption prevails throughout the hierarchy of the PDS officials. There are also complaints of large-scale duplicates and ghost ration cards, which the uniqueness of Aadhaar can help to reduce. If the entire database of ration cards is seeded with Aadhaar numbers, one can eliminate the bogus ration cards. Authentication helps in ensuring delivery of ration to the beneficiaries of the PDS.

The PDS owner has a biometric authentication device connected to the entitlement database at the back end. This database includes details of all the family members on the ration card. Once a customer provides his ration card number, the shop owner selects the individual who has come to make the purchase, fills in the requested items and authenticates the beneficiary's identity through his fingerprint or iris scan. At the back end, the authentication packet is transmitted to UIDAI, which confirms the ID of the person, after which the PDS back end allows the transaction of ration purchase to take place. This is a foolproof system where all the stakeholders: the PDS shop owner, the beneficiary and the government are participants. The transaction is traceable and transparent. There are less complaints now as there is a digital proof of this transaction.

The most important advantage of this system is portability. As the entitlements, identities and inventory of the ration at various FPS is online, one can manage delivery choices more efficiently. Many delivery systems have now used UIDAI's authentication services, thereby validating the assumptions, the design and its implementation. Aadhaar, therefore, equips any resident, especially the poor who often don't have alternative means, to access formal services. It is due to Aadhaar that we can realize the vision of One Nation One Card.

Birth of JAM

Aadhaar has also been used as a financial address and leveraged for DBT in several domains such as LPG (PAHAL), scholarships and social security pensions (NSAP). The eKYC has also been used for opening bank accounts under the Pradhan Mantri Jan-Dhan Yojana[190] (referred to as Jan-Dhan here) and for getting mobile SIMs. The impact of these systems is profoundly enabling and has brought a positive change in the lives of hundreds of millions.

By linking their mobile number with Aadhaar, the poor can conduct banking transactions or access government services that are offered online. The authentication service offered by UIDAI becomes extremely useful, both at the time of establishing relationship with a bank (i.e., opening of bank account) and while executing any transactions, the equivalent of identification by matching signature.

On 27 February 2015, the *Economic Survey*[191] gave a concept that would change, and continues to change, the landscape of welfare and poverty alleviation in this country through Jan-Dhan, Aadhaar and Mobile or the JAM trinity. It presented a cohesive and moving argument that the trio of a low-cost bank account, unique digital identity and ubiquitous communication device could be a big boost in the fight against poverty. The impact of this trinity is profoundly enabling.

A study presented in the *Economic Survey*, 2015, showed that price subsidies are often regressive, distorting markets in ways that ultimately hurt the poor. The government provides price subsidies for commodities and services—sometimes both at the production and distribution ends—such as rice, wheat, pulses, sugar, kerosene, LPG, naphtha, water, electricity, diesel, fertilizer, iron ore and railways. The survey provided two specific examples of price subsidies for kerosene and rice in the PDS that do not work the way they should.

Leakages are universal and increase with the size of PDS

---

[190]Pradhan Mantri Jan-Dhan Yojana. Available at: https://pmjdy.gov.in/. Last accessed on 15 June 2020.
[191]*Economic Survey*, Chapter 3, 'Wiping Every Tear from Every Eye: The JAM number trinity solution,' 2015. Available at: https://www.indiabudget.gov.in/budget2015-2016/es2014-15/echapvol1-03.pdf. Last accessed on 19 May 2020.

allocations. For instance, the survey states that only about 46 per cent of the subsidized kerosene is used by the poor, while the total PDS allocation of kerosene exceeds its total consumption! Similarly, it indicates significant leakages in food distribution through PDS, the cost of which could be as high as ₹19,000 crore (₹190 billion) per annum on wheat and rice alone. This is quite believable, being less than 15 per cent of the whopping ₹1,29,000 crore (₹1.29 trillion) in subsidies that the government paid for these two commodities.[192]

Unfortunately, the subsidies that miss their target escape to those who are better off, and whose consumption of the subsidized items is more—fertilizers and electricity being two prime examples. The highly subsidized PDS rates for commodities such as rice and wheat create the incentive to sell these into a thriving black market.

Taking the subsidy out of the price of goods and channelling them separately and directly into the hands of the targeted population foils the black marketeers, as was done in the case of LPG subsidy through the PAHAL programme. It was enabled by the direct transfer of subsidies into the beneficiary's Aadhaar-linked Jan-Dhan bank account, with a permanent advance of one cylinder's subsidy to reduce the burden. It was initially launched on 1 June 2013 and covered 291 districts. A modified scheme was relaunched in 54 districts on 15 November 2014 in the first phase and then the rest of the country on 1 January 2015.[193]

The theme of leveraging Jan-Dhan accounts, Aadhaar and Mobile (JAM) was reprised the following year in the *Economic Survey 2015–16*.[194] It notes that in that period, 'Jan-Dhan and Aadhaar deepened their coverage at an astonishing rate—creating between

[192]For the procurement of rice, the Food Corporation of India (FCI) pays the minimum support price (MSP) of ₹14 per kg and sells the rice at ₹1.9 per kg to its network of FPSs, who in turn, sell it to entitled consumers at ₹2 per kg. The difference is the subsidy paid for by the government.
[193]PAHAL DBTL Consumers Scheme, Ministry of Petroleum and Natural Gas. Available at: http://petroleum.nic.in/dbt/whatisdbtl.html. Last accessed on 9 May 2020.
[194]*Economic Survey*, Chapter 3, 'Spreading JAM across India's Economy', Available at: https://www.indiabudget.gov.in/budget2016-2017/es2015-16/echapvol1-03.pdf. Last accessed on 19 May 2020.

two million and four million accounts per week respectively' and the 'most promising targets for JAM are fertilizer subsidies and within-government fund transfers—areas under significant central government control—with substantial potential for fiscal savings.'

Till the end of February 2019, the government's move towards DBT has enabled the transfer of more than ₹6,80,056 crore (₹6.80 trillion) through as many as 440 schemes under 55 ministries. This has saved an estimated ₹1,20,469 crore (₹1.20 trillion).[195] Thus, Aadhaar streamlines and cleans up the delivery process, prevents denial of benefits out of malfeasance, empowers the poor to access banking, insurance or other services, and brings the efficiency and convenience of digital systems.

Uses of JAM

Aadhaar is a special kind of public good. It has been created with public money and its eKYC and authentication services are non-rivalrous. This means neither private interests nor 'unreasonable' government restrictions ought to prevent residents or legal Indian entities from making use of this infrastructure.

This isn't as radical as it sounds. Aadhaar is a public infrastructure and available for public use in a non-discriminatory manner to anyone who is willing and able to pay a reasonable charge to cover the marginal cost of the consumed product or service. For instance, the government may build a highway out of taxpayer money or through a public–private partnership agreement and then make it available to everyone without charge or for a small fee. In the normal course, it may not declare that the highway is for use of government vehicles alone. (This is where the decision of the Supreme Court had created the limitation.) Thus, the same infrastructure that enables the efficient delivery of public services, could enable low-cost delivery of private services as well. For instance, a new

---

[195]Surabhi, 'Direct Benefit Transfer of Subsidies Crosses ₹3-lakh-cr Mark in 2018-19', *The Hindu BusinessLine*, 22 April 2019. Available at: https://www.thehindubusinessline. com/economy/direct-benefit-transfer-of-subsidies-crosses-3-lakh-cr-mark-in-2018-19/ article26913247.ece. Last accessed on 8 May 2020.

telecom operator who entered the market was able to climb to 100 million customers in five months because Aadhaar-based eKYC made the process simple and economical. As other telcos responded to the competition, in 2016–17, the consumption of mobile data in India rose nine times, making the world sit up and take notice.

Another example is what happened in the mutual funds domain. Given the reduced cost of selling mutual funds, low-income group customers were also targeted to move away from traditional investment assets such as gold and property to better-performing financial market instruments. The advertising campaign itself underwent a change in tone from the fast-paced 'Mutual fund investments are subject to market risks. Please read the offer document carefully before investing' in English to a more mass-market *'Mutual funds sahi hai* (Mutual funds are good).' The increased inflows from domestic investors have helped buoy the market and decrease volatility from flighty foreign capital.

The private sector fulfils functions that the state is not meant to provide. For example, timely and low-cost access to credit is necessary for people to grow their incomes and stabilize their financial lives. Without JAM, the poor would be financially excluded. With only JAM, they may have a bank account, but not necessarily all the benefits of banking.

The onslaught of COVID-19 has revealed the strengths and weaknesses of governments and public infrastructure across the globe. The economic ramifications have wreaked havoc for the most vulnerable among us. In such daunting circumstances, policymakers across the world have explored using a range of instruments to blunt the adverse impact of the pandemic—from economic packages for industry to direct benefit transfers (DBTs) for citizens, all options have been on the table.

The JAM trinity has proved to be a silver lining for India in these grim times. The average usage of the Aadhaar-enabled payment system (AEPS) doubled to 11.3 million per day during the COVID-19 lockdown that started on 25 March 2020, as beneficiaries of the ₹1.7 lakh crore (₹1.7 trillion) Pradhan Mantri Garib Kalyan Yojana (PMGKY) tapped the system to withdraw money transferred by the government to their accounts.[196] The government has so far transferred ₹32,300 crore (₹323 billion) to bank accounts of 340 million beneficiaries since the launch of the ₹1.7 lakh crore (₹1.7 trillion) welfare scheme on 26 March 2020 by Finance Minister Nirmala Sitharaman to protect the poor from the impact of the lockdown.

While it is often said that one must not waste a crisis, preparedness to deal with a crisis can go unnoticed with many other imperatives at hand. Thanks to JAM, India has arguably the most sophisticated benefits transfer infrastructure in the world. The pandemic only proved its worthiness.

IndiaStack for Low-Cost Services

Today, India has a full suite of products based on Aadhaar-authentication services for a paperless, cashless, presence-less, transparent and consent-based architecture, what is commonly described as IndiaStack. This is a set of open APIs that enables services to be constructed from building blocks of simpler services, tested and deployed at scale. Using the IndiaStack's Open APIs, Capital Float, a digital finance company that provides collateral free unsecured business loans, could authenticate an individual, open an account and collect data from sources, such as the user's phone itself, to determine whether he or she could be given a small loan. The amazing thing is that all of it could be done in less than a minute!

IndiaStack has enabled the private sector to deliver low-cost services, and provide people access to a wider variety of financial services and products that suit their needs. The economic value created by everyone using an open platform far exceeds the cost of the underlying platform itself. Take the example of TCP/IP protocol that came to be widely adopted for networking computers.

[196]Rajeev Jayaswal, 'Use of Aadhaar-enabled Payment Service Doubles,' *The Hindustan Times*, 5 May 2020. Available at: https://www.hindustantimes.com/india-news/use-of-aadhaar-enabled-payment-service-doubles/story-q3ZaAZ60qPC5tnX3NKhOWL.html. Last accessed on 1 June 2020.

It provided basic services for dissimilar networks and computer systems to interoperate, resulting in all the magnificent services that comprise the internet today! Similarly, Aadhaar is a platform that can be used to fulfil a basic function: to uniquely recognize and authenticate individual residents of India, which has already enabled much economic and social empowerment in a wide variety of fields. But its true potential knows no boundaries.

Quantifying Benefits

In January 2019, then finance minister, Late Arun Jaitley, said that government estimates put Aadhaar savings at ₹90,000 crore (₹900 billion) till March 2018.[197] On this Facebook post, he also quoted the World Bank's *Digital Dividend Report* that India can save an estimated ₹77,000 crore (₹770 billion) every year using Aadhaar. These savings are huge—an order of magnitude bigger than the expenditure that has been incurred on Aadhaar. But being estimates, disbelievers choose to cavil at the figures without making an effort to collect more accurate data themselves.

I would argue that actual savings due to Aadhaar are an order of magnitude 'larger' than what the GoI or the World Bank have estimated. Consider this: what happens if food subsidy of ₹100, which was meant for an undernourished boy or girl, is captured by someone not entitled to it? Aadhaar savings have been estimated on the basis of prevention of such loss. But researchers have determined[198] that malnutrition in children (or mothers that bear them) leads to serious health, educational and economic consequences throughout their later life.

What are the consequences of government subsidies (supplemented with proper education) not reaching the intended

---

[197] Arun Jaitley, 'Benefits of the Aadhaar – where it stands today,' Facebook, 6 January 2019. Available at: https://www.facebook.com/notes/arun-jaitley/benefits-of-the-aadhaar-where-it-stands-today/940008419521040/

[198] Cesar G. Victora, et al., 'Maternal and Child Undernutrition: Consequences for adult health and human capital,' *The Lancet,* Vol. 371 (9609), 17 January 2008, pp. 340–57. Available at: https://www.thelancet.com/article/S0140-6736(07)61692-4/fulltext. Last accessed on 8 May 2020.

population? This is what the Nobel Prize-winning economists Abhijit Banerjee and Esther Duflo, stated in 2011:

> ... the study of the long-term effect of deworming children in Kenya, concluded that being dewormed for two years instead of one (and hence being better nourished for two years instead of one) would lead to a lifetime income gain of $3,269 USD PPP. Small differences in investments in childhood nutrition (in Kenya, deworming costs $1.36 USD PPP per year; in India, a packet of iodized salt sells for $0.62 USD PPP; in Indonesia, fortified fish sauce costs $7 USD PPP per year) make a huge difference later on.[199]

The multiplier effect of reaching targeted benefits into the right hands of intended recipients could be hundred or thousand times the money value of the benefit. In this country, large sections of population are excluded by poverty, malnutrition, lack of education and access to modern instruments of economic empowerment from partaking in the benefits of human progress. An improvement in targeted deliveries in India could thus be the difference between the country reaching the potential that the framers of the Constitution envisioned for it—or not. That value would translate into trillions of dollars, far above a billion or so spent on Aadhaar.

Digital Consent Artefact

In the twenty-first century, data has emerged as a key asset for development, both for the public and private sectors. Facebook, Google, WhatsApp, Amazon—some of the largest twenty-first-century corporates—have built their fortunes from the data they collect. This creates tension between individuals, data providers and data consumers. The individual, to whom the data pertains, must have control over who gets to see it, in which form and for what purpose—but doesn't.

---

[199]Abhijit V. Banerjee and Esther Duflo, *Poor Economics: A radical rethinking of the way to fight global poverty.* Noida: Random House, India, 2011. Print.

Though it has been targeted for violation of individual privacy, ironically it was Aadhaar that pioneered a safe and consent-based data regime. UIDAI was the first entity to adopt a minimalist data-collection design at a time when both governments and corporates were indiscriminately hunting data. It championed consent-based data sharing. And it was the first to frame rules, however elementary, on data usage. Finally, by introducing server-to-server data transfer, fully encrypted end to end, it secured data like never before.

The digital consent artefact has been developed using some of these characteristics of Aadhaar, for use by the public and the private sector.[200] It is as yet a nascent component of IndiaStack, comprising an electronic consent architecture built around four pillars: the individual or the owner of the data; the consent collector, which is a regulated agency; data providers such as banks, telcos, hospitals, etc., and data consumers such as banks, credit providers, etc.

This data consent artefact creates a data exchange in which those who need the data can obtain it from those willing to provide it, but with the mediation of a data fiduciary that acts strictly in the interest of the individual, the owner of the data. For example, when an individual seeks a small loan, the lender would naturally want to determine whether the seeker is an acceptable credit risk before making the agreement. The information that the creditor needs may reside in myriad online repositories that either obtained it from the applicant or created it as a by-product of their own service offering. The individual's credit repayment history, social standing or consumption pattern, as in payment of electricity or telephone bills, could all be part of the data used in the process. The consent artefact enables the individual as the loan applicant in this case to share such information to support his own application process.

The individual can shop for the best loan terms and lenders can mine big data to make their business decisions, all in minutes, without the need for physical contact, and most importantly, without

---

[200]Electronic Consent Framework: Technology Specifications. Available at: http://dla. gov.in/sites/default/files/pdf/MeitY-Consent-Tech-Framework%20v1.1.pdf. Last accessed on 8 May 2020.

the individual losing autonomy over her data. Like a lubricant that eases the movement of the smallest of parts, and thus makes it possible to put a huge machine in motion, the consent artefact, by removing friction from transactions could crank up the entire economy!

The consent artefact is a modern, privacy-enabling framework for sharing data. It is consistent with current legal frameworks and compliant with the Information Technology Act. It is user-centric, which means individuals have full control over their data that can only be shared after obtaining consent through Aadhaar authentication. The use of Aadhaar makes data sharing auditable and non-repudiable, as authorization logs prove both consent and authorization. Deployment of Aadhaar also builds trust as digital signature is a prerequisite to share data at every stage.

These schemes and programmes are just the tip of the iceberg. India has not yet fully unlocked the power of JAM. DBT will reach an estimated 75 per cent of all central government subsidies at approximately ₹4,00,000 crore (₹4 trillion) in the financial year 2019.[201] But ₹3,00,000 crore (₹3 trillion) of state subsidies still remain. Citizen's data is still not fully on DigiLocker, which is a flourishing repository of various kinds of official documents such as degrees, certificates, records and transactions. Digitization of schemes will also help.

Opportunities in the Aadhaar Ecosystem

Like infinitely new models could be created by bolting together plates, angles, gears and wheels in a Meccano set, Aadhaar and its related artefacts provide the building blocks for more complex solutions. What you do need to provide is imagination in solving the problem. As Nandan often used to say: Aadhaar is like a door that opens other doors.

---

[201]'Big Year Ahead for Direct Benefit Transfer: Over 75% subsidies could be paid by DBT in FY19,' *The Financial Express*, 5 March 2018. Available at: https://www.financialexpress.com/opinion/big-year-ahead-for-direct-benefit-transfer-over-75-subsidies-could-be-paid-by-dbt-in-fy19/1087371/. Last accessed on 8 May 2020.

Let's select a problem at random and see how we could build a prototype solution. The actual solution would, of course, require more engineering effort and validation, but the process can be illustrated relatively easily in a few paragraphs.

If Aadhaar is the foundation of citizens' empowerment, their needs are first and foremost. These needs are not all met by the government, nor is it necessary.

People have a right to unrestricted movement anywhere in the country. But when they exercise that choice, their service providers need to be notified of the change in the address. That's a problem to which the government or private enterprise could provide a solution.

The resident could register their Aadhaar number (or to be safer, their virtual Aadhaar number) and address with a record management service (RMS). The RMS would generate a token or several tokens that represent the address, which you can share with the service providers you use. When you change the address, anybody who can present the address token to the RMS is provided with the latest address.[202]

You can revoke access to one or more agencies at any time, and your RMS would know who has your updated address and who doesn't. There is no loss of privacy in this arrangement because your RMS acts on your express wish, whether as a standing mandate or on case-by-case basis. There is also no profiling possible because you only share a virtual Aadhaar with the RMS and, further tokens that it generates need only be used once for each mandate. This mechanism can notify a change in telephone number to your friends; change in ownership of an asset such as a motor car to the insurance company; change in service provider (a new insurance company or plan modifications to your health-service provider, or vice versa); change of telephone company to your bank for bill payments, etc.

Indeed, some functions that the government provided in the past, such as registration of deeds, could also be entrusted to private enterprise with the help of Aadhaar for identity and a blockchain for

---

[202]It could easily be arranged for the address to be communicated on your behalf as a push notification too.

immutability of records. Or consider the reputation of an individual, which itself is an asset. If reputational scores could be attached to the identity of an individual, it may encourage people to invest in their reputation much like the Uber driver who would rather not displease a customer and get less than a five-star rating. Here, I'm not suggesting government-mandated scores, but the possibility that scores could be created in a variety of ways by different agencies, all with the explicit knowledge of the individual and with her willing participation.

The reader would note that not just Aadhaar, but post-Aadhaar applications too have impacted the lives of ordinary citizens, often providing frugal and scalable solutions especially suited for India. These are case studies for other countries that wish to solve similar problems, but without the constraints of old-school solutions that continue to be the mainstay of even the advanced economies.

These solutions outlined above are offered only as quick improvisations to illustrate how easily new applications could be conceived and engineered. I would urge even the sceptics to use their imagination and intellectual prowess to invent their own solutions to the problems closest to their hearts.

As that song from the 1970s says, 'You Ain't Seen Nothing Yet'!

EPILOGUE

Aadhaar could be developed and implemented successfully because we had a precisely defined problem to work upon: how would you provide a unique identity to each Indian resident, if they ask for it?

When everyone, from the CEO to the janitor, knows what the organization's objective is, the decision-making becomes easier; whether it is to converge upon the broad solution or to make hundreds of smaller decisions that need to be taken in its implementation. In the case of UIDAI, even the man in the street knew what the organization aimed to provide. 'Unique Identification' was part of its name!

However, to achieve rigour in action, it is of equal importance to untangle and cast away what you do not promise. For instance, Aadhaar promised no proof of citizenship and no eligibility to benefits. It collected no data inessential to providing identity and offered no services other than a simple authentication after the Aadhaar number is issued.

By isolating a single function, UIDAI could focus on doing just that in a frugal manner, avoid the legal and policy pitfalls and make the solution robust and scalable. Other functions could then be built upon the core functionality of uniquely identifying a person that Aadhaar provided. Let me illustrate how robustness springs from fully solving the core problem.

Making Processes Simpler

Aadhaar was meant to provide inclusion, even for those who did not have a proper identity document.[203] As we chose a robust biometric deduplication process, nobody could obtain two Aadhaar numbers.

---

[203]Swetha Totapally, Petra Sonderegger, Priti Rao, Jasper Gosselt and Gaurav Gupta, '49% of Residents in a Recent Survey Were Found to Have Used Aadhaar to Access One or More Services for the Very First Time!' *State of Aadhaar Report 2019*. Dalberg, 2019.

That made it easier to accept weaker forms of identity documents for enrolments such as a ration card or a simple letter from the local representative. As Aadhaar was not linked to any benefit, a resident had no incentive to fabricate basic details about herself because the Aadhaar they got would be linked to their own biometrics. They could get an error corrected (say, age) or a detail modified (such as their mobile number or address). They had no incentive to be untruthful because a lie that is expedient in one situation would prove damaging in another.

This fact allowed us to make the enrolment process convenient for the ordinary resident, who did not have to answer endless questions or provide documentary proof for everything. He or she could be relied upon to tell the truth with just basic verification. It also allowed UIDAI to build standard, technology-driven processes through which non-government employees could act as enrolment agents. Again, as payment to enrolment agencies was linked to the outcome—i.e., successful generation of the Aadhaar number—it kept the agents diligent and careful.

It is only by making processes simpler and aligning the incentives to a single objective that UIDAI could cover most of the population in a few years and at one-fifteenth the cost estimate that an 'outside' expert provided to the Parliament Standing Committee on Finance.

In December 2011, this committee, after hearing many experts, determined that:

> The UID scheme has been conceptualized with no clarity of purpose and leaving many things to be sorted out during the course of its implementation; and is being implemented in a directionless way with a lot of confusion.[204]

The parliamentarians appeared to have been led to doubt the technical feasibility and the cost estimates by the self-proclaimed experts that testified before it. Unfortunately, the experts did not seek answers to questions they may have had, but chose to express

---

[204]Parliamentary Standing Committee on Finance, *Report on the National Identification Authority of India Bill, 2010*, 13 December 2011.

their doubts as incontestable conclusions.

However, the clarity of vision in the Strategy Document helped UIDAI officers find ways to overcome the innumerable challenges that came up. They knew what could not be compromised at all and everything else became fodder in creative problem-solving. This is what allowed us to build a team of people from disparate backgrounds, where everyone contributed, everyone learnt and none experienced the usual constraints.

We steered clear of proprietary solutions and integrated market dynamics in every aspect of work. Companies were required to compete for the deduplication pie, suppliers of hardware or software had to ensure interoperability and quality-cum-efficiency became the sole competitive advantage in enrolment work.

We had a one-point agenda. As we were solving a common problem of all domains, we were able to interact freely with all stakeholders. To a large extent, it reduced the friction usually arising out of turf issues in government systems. We could listen to their concerns and address them right from the design stage. Of course, we could not win over those who, per se, had an ideological opposition to unique identities.

A project that touches so many lives and impacts systems that have been in existence over a long time cannot succeed without political support. It is tempting to argue that political support can be obtained for a project because it has clarity of purpose, a good design and effective implementation. These are necessary conditions, but they are by no means sufficient. UIDAI was fortunate to have the support of a stalwart like Shri Pranab Mukherjee, then finance minister and the chairman of EGoM, who was looking into the issue of creating unique identity since 2006. Aadhaar also enjoyed bipartisan support that Nandan was able to create by tirelessly engaging with top leaders at the Centre and the states. Later, it was lucky to have the support of Shri Modi as the PM. Once convinced of its merit, he did not hesitate to take a position contrary to his party's earlier stand and took Aadhaar to new heights, fully leveraging it to the benefit of the poor and the marginalized.

While technology undergirds the project, its philosophical

underpinnings have an important influence too. Aadhaar solved one hard problem, that of identity, and did so in a manner that other systems could incorporate the solution at virtually no extra cost.

Thus, the Income Tax department could stop worrying about one person fraudulently obtaining two PAN cards.[205] The Income Tax Act made it illegal for anybody to do so; UIDAI made it impossible. Similarly, if a driving licence or passport or the recipient of subsidy in LPG is linked to Aadhaar, the same person cannot get two documents for the same purpose or twice the subsidy.

Solve once and solve for everyone. This is the same philosophy in DigiLocker, which provides a single repository of digitally signed documents—university degree certificate or driving licence—issued in one person's name, hosted in one place, secured and stored in perpetuity and conveniently accessed by the user.

NPCI has likewise created a payment system, UPI, which solves the singular problem of digital transfer of money from one bank account to another. Users can create a virtual address and link it to their accounts (as many virtual addresses as they need, one for each account). They may now transact amongst themselves using any of half a dozen popular apps. This payment solution has become so popular that it now averages a billion transactions per month.[206] In three years since its launch, UPI is the big daddy of payments in India because it exceeds all other forms of electronic payments: credit cards, debit cards, net banking, wallets or other such mechanisms. Recently, Google recommended to the US's Federal Reserve to develop its real-time gross settlement service in line with India's UPI.[207]

This idea of isolating and solving one problem is especially

---

[205]Permanent Account Number (PAN) is a 10-character alphanumeric identifier, issued by the Indian Income Tax Department, under Section 139A of the Income Tax Act
[206]'NPCI, UPI Product Statistics, 1,148.36 million transactions with aggregate money value of ₹1,91,359.94 crores in October 2019.' Available at: https://www.npci.org.in/product-statistics/upi-product-statistics. Last accessed on 8 May 2020.
[207]IANS, 'Google Wants US Federal Reserve to Follow India's UPI Example,' *The Economic Times*, 15 December 2019. Available at: https://economictimes.indiatimes.com/news/economy/policy/google-wants-us-fed-reserve-to-follow-indias-upi-example/articleshow/72611611.cms?from=mdr. Last accessed on 8 May 2020.

powerful in the digital world because the resulting functionality can often be offered as a set of API calls. Think of these APIs as a way for machines to interact with each other over the internet. Want a sentence translated into another language? Make an API call to a server that can do it for you, for a charge or free. Want to get the format of a document changed from .pdf to .docx? Want to get a voice clip converted to text? Bingo, make an API call!

What this idea can do for governance is transformative. A fair bit of services in the modern economy are already delivered via the internet, whether by government or private enterprise. The government is often also required to make and enforce rules, such as how insurance must settle a claim for medical expenses or hospitals must safeguard and yet share the patient's personal health records upon request by the individual himself/herself or his/her family. If you isolate a function, such as obtaining consent about sharing of medical history, the government could enforce regulations upon the registered provider of such services. The data could be obtained, protected and transferred in accordance with this consent mediated by service requests between the citizen, the government and third-party service providers.

'Make in India' Story

You can see the beginnings of such implementations in the functions that are hosted or supported by the government today: want to confirm the name of the person who offers his Aadhaar number and a finger scan? Make an API call through a registered provider and UIDAI will 'authenticate'. Want to digitally sign your document? Fetch a copy of your driving licence? Fetch a utility bill? Make a payment on your behalf? There are APIs for all these: eSign, DigiLocker, Bharat Bill Payment System (BBPS) and UPI, respectively. Converting government services into a federated set of essential, atomic tasks that are delivered as interoperable functions is what will deliver the promise of 'minimum government and maximum governance'.

One final word about the realization that is emerging today:

data is precious because it is the source of all insights. But no one has an answer to the conundrum about how to share and protect it at the same time.

Data about traffic congestion, your specific location and the roads that lead to your destination can ease your journey. This data, however, cannot be collected and used without your consent (data right), a fair compensation (share in the value of that data) and a guarantee that it won't be used for any other purpose than what you agreed to (privacy protection). Such data collection is becoming all-pervasive: voice samples for training speech artificial intelligence (AI), the songs you play for music recommendation engines, your picture for improving facial recognition or your medical history for research into disease processes.

Large technology companies now collect and analyse every mouse click or pause as you scroll through your social media stream, use an app, make a search or even directly type a website address into the browser. This has generated a mind-boggling capability to not only understand but influence people as individuals. It's a massive surveillance infrastructure for concentrating profits and economic power in a few hands. It is equally worrisome that this infrastructure has the potential for abuse in manipulating the thoughts of people at scale!

How would the individual's right to life and liberty be protected without adequately safeguarding their data and without them sharing in the economic benefits released from its use? There are no easy answers to these questions. However, policymakers, legal brains and technologists can profit by analysing trade-offs in each indivisible, essential function, just as UIDAI did, for identity and authentication.

There is a huge opportunity in this for India. We have shown the world how identity projects could be done and how other services can be built upon the identity infrastructure. It is truly a 'Make in India' story because the solutions were designed and implemented here.

Bringing into play the latest addition to this arsenal, the Electronic Consent Artefact, could allow us to hold the data in a federated ecosystem of entities. So, protected from unbridled access, use of the data can be governed by the explicit consent of the individual to whom it pertains, restoring the balance of power between the individual and other powerful actors, whether the state or large private enterprises.

Building one service, one functionality or one solution at a time, we could prevent concentration of data in the hands of a few corporations in what otherwise looks like a dystopian future. These ideas, starting from the fundamental building block of identity, may be simple, but building the architecture and products to operationalize them is a valuable intellectual property right (IPR).

India may have quietly started a movement that could shape the world of tomorrow. We should not hesitate to take a leadership role in this movement just because nobody in the world is currently doing it or considers it their problem to solve. It is only by doing what India needs and sharing our solutions with the world can we earn our rightful place in the world order.

That may well be the most instructive use of the story told here.

ACKNOWLEDGEMENTS

I had never dreamt of writing a book. It wasn't something that science students, who could afford to be dull so long as they were precise and factual, are equipped to do. My career experience, too, was limited to short notes on files and an occasional report of no literary merit whatsoever.

When Nandan was writing *Rebooting India: Realizing a Billion Aspirations*, his co-author, Viral would ask for my inputs on the Aadhaar project. I provided copious notes in response. I was under the impression that the book was about the Aadhaar experience. However, when it came out, I found that the book covered many subjects and ideas besides Aadhaar. While Nandan had made broad observations and captured key lessons from the Aadhaar experience, there were many aspects of project execution which were not included.

I thought that story needed to be told and shared the idea of a book with Nandan, who encouraged me to go ahead. The rest, as they say, was a lot of labour.

I approached erstwhile colleagues and asked them to share their perspective, which many did. I am grateful to them and have tried to credit their inputs in the book itself. However, I would also like to mention a few of the names here: Ashok Pal Singh, Rajesh Mashruwala, D. Subhalakshmi, Raju Rajagopal, Pramod Varma, Naman Pugalia, Kabir Shetty and Asif Iqbal. It was a privilege to have worked with them and my great fortune to have had their encouragement and support so many years later for a book like this. Alok Shukla, who had worked with me in UIDAI and continues to work there, was invaluable in providing references to events, dates and documents. Without these efforts, the book may have only remained a wish.

If you are disciplined enough to put in the effort, progress is a natural outcome. Soon the draft started to emerge, but I was unhappy with what I saw. I shared those chapters with Ashok Pal

Singh, Naman Pugalia and Arvind Verma. Their suggestions and corrections helped to bring out the story that otherwise threatened to drown in the early text.

Finally, entered Sunil Bajpai, my colleague in TRAI. When I mentioned my problem to him, he agreed to help out. Since then, this project acquired the required momentum and speed. Sunil has revised and improved the document significantly through many iterations. He is truly my co-author.

I wasn't able to do justice to all the inputs from my erstwhile colleagues and now friends due to constraints of space. My apologies to them.

I am thankful to Rupa Publications for its support and the flexibility in timelines extended to me. The commissioning and editorial teams at Rupa have also provided excellent editorial support, and their firmness and inflexibility were very helpful too.

Kalpana, my wife, has been an equal partner in 37 years of adventure that my well-wishers often considered 'whimsy bordering on foolishness'. Her strength and support gave me courage to pursue the life that promised the greatest fulfilment. No words can acknowledge her contribution or express what she means to me.

Lastly, let me express my sincere gratitude to all my colleagues in UIDAI and others who helped make this project a success. It is their hard work and commitment that the unique ID project wasn't thrown in the dustbin of failed projects and forgotten like many others. I owe it to them that I could think of writing this book.

I share a strong belief with Nandan that we can solve many of India's hard problems using technology. That's what impelled me to write the book despite the initial aversion for such exertions. I think we should celebrate Aadhaar and continue to build more and more technology products, which will make a positive impact in the governance of our country and the lives of our countrymen.

ABBREVIATIONS

ADB	Asian Development Bank
AEBAS	Aadhaar-enabled Biometric Attendance System
AePDS	Aadhaar-enabled Public Distribution System
AEPS	Aadhaar-enabled Payment System
APB	Aadhaar Payment Bridge
API	Application Programming Interface
ASDMA	Application Software Development, Maintenance and Support Services Agency
AUA	Authentication User Agency
BATF	Bangalore Agenda Task Force
BBPS	Bharat Bill Payment System
BDO	Block Development Officer
BPL	Below poverty Line
BSP	Biometrics Solutions Provider
CAG	Comptroller and Auditor General of India
CBP	Customs and Border Protection
CCT	Conditional Cash Transfer
CCUIDAI	Cabinet Committee on UIDAI
CE	Chief Engineer
CIDR	Central Identities Data Repository
COAI	Cellular Operators Association of India
CPGRAMS	Centralized Public Grievance Redress and Monitoring System
CR	Change Request
CSC	Common Service Centre
CSO	Civil Society Organization

CVC	Central Vigilance Commissioner
DBT	Direct Benefit Transfer
DC	Deputy Commissioner
DDG	Deputy Director General
DEA	Department of Economic Affairs
DFS	Department of Financial Services
DG	Director General
DM	District Magistrate
DoT	Department of Telecommunications
DPA	Data Protection Authority
DSC	Digital Signature Certificate
DTO	District Transport Office
EGoM	Empowered Group of Ministers
EPIC	Elector's Photo Identity Card
EPOS	Electronic Point of Sale
EVMs	Electronic Voting Machines
FDM	Federated Data Model
FIU	Financial Intelligence Unit
FMR	False Match Rate
FPIR	False Positive Identification Rate
FPS	Fair price shop
FRR	False Reject Rate
GO	Government Officer
GoI	Government of India
GPF	General Provident Fund
IAAS	Indian Audit and Accounts Service
ICA	Immigration & Checkpoints Authority
ICT	Information and Communication Technology
IDAS	Indian Defence Accounts Service
IIAS	Indian Institute of Advanced Studies

IIT	Indian Institute of Technology
IMF	International Monetary Fund
IOC	Indian Oil Corporation
IPC	Indian Penal Code
IPoS	Indian Postal Service
IPS	Indian Police Service
IRDA	Insurance Regulatory and Development Authority
ITS	Indian Telecom Service
ITU	International Telecommunication Union
JAM	Jan Dhan-Aadhaar-Mobile
KYR	Know Your Resident
KYR+	Know Your Resident Plus
LIC	Life Insurance Corporation of India
MGNREGA	Mahatma Gandhi National Rural Employment Guarantee Scheme
MHA	Ministry of Home Affairs
MIS	Management Information System
MISM	Masters in Information Systems Management
MIT	Massachusetts Institute of Technology
MKSS	Mazdoor Kisan Sangram Samiti
MNIC	Multipurpose National Identity Cards
MoF	Ministry of Finance
NAC	National Advisory Council
NCMP	National Common Minimum Programme
NDA	National Democratic Alliance
NIC	National Informatics Centre
NIPFP	National Institute of Public Finance and Policy
NIRD	National Institute of Rural Development
NISG	National Institute for Smart Government

NPCI	National Payments Corporation of India
NPR	National Population Register
NREP	National Rural Employment Programme
NRIC	National Register of Indian Citizens
NSAP	National Social Assistance Programme
NYSE	New York Stock Exchange
OPT	Optional Practical Training
ORS	Online Registration System
OSN	Online Social Networks
PAHAL	Pratyaksh Hastantarit Labh
PAN	Permanent Account Number
PbD	Privacy by Design
PDS	Public Distribution System
PETs	Privacy-enhancing technologies
PF	Provident Fund
PFRDA	Pension Fund Regulatory and Development Authority
PG	Public Grievances
PIL	Public interest litigation
PIN	Personal identification number
PMO	Prime Minister's Office
PMU	Project Monitoring Unit
PNR	Passenger Name Record
PoC	Proof of Concept
PPP	Public–Private Partnership
PSRs	Private-sector resources
PSU	Public-sector Undertaking
PwC	PricewaterhouseCoopers
PWD	Public Works Department
QA	Quality Assurance

RBI	Reserve Bank of India
RC	Registration Certificate
RFID	Radio-Frequency Identification
RGI	The Registrar General of India
RMS	Record Management Service
RTI	Right to Information (RTI) Act
SE	Superintending Engineer
SEBI	Securities and Exchange Board of India
SFinGe	Synthetic Fingerprint Generator
SFTP	Secure File Transfer Protocol
SOP	Standard Operating Procedure
SRS	Software Requirement Specifications
SSN	Social Security Number
STC	State Transport Commissioner
TDU	Technology Development Unit
TRAI	Telecom Regulatory Authority of India
UID	Unique identification
UIDAI	Unique Identificati on Authority of India
UK ICO	United Kingdom's (UK) Information Commissioner's Office (ICO)
UPI	Unified Payments Interface
UPSC	Union Public Service Commission

INDEX